NOKIA'S SMARTPHONE PROBLEM

PROBLEM

THE END OF AN ICON?

Majeed Ahmad

SMARTPHONE CHRONICLE

This publication is designed to provide accurate and author-itative information in regard to the subject matter covered. It is sold with the understanding that the publisher is not engaged in rendering professional services. The advice and strategies contained herein may not be suitable for your sit-uation. You should consult with a professional where appro-priate. Neither the publisher nor author shall be liable for any loss of profit or any other commercial damages, includ-ing but not limited to special, incidental, consequential, or other damages.

First edition published on May 6, 2013.

ISBN-10: 1482091232
ISBN-13: 9781482091236
Library of Congress Control Number: 2013902207
CreateSpace Independent Publishing Platform
North Charleston, South Carolina

CONTENTS

PROLOGUE

"About every five years, like clockwork, the nature of technology and the nature of the market accumulate enough change that it feels like a disruption."

— Ray Ozzie, inventor of Lotus Notes and former chief software architect at Microsoft

Motorola Inc. rode the crest of the analog mobile phone wave during the 1990s until a Finnish wireless upstart came along. Nokia Oyj ended Motorola's reign, and despite years of reorganization and the introduction of some very successful products, Motorola eventually became a faint shadow of its glorious past. The modern digital mobile phones were purely devices for conversation and text messages. Nokia was the company which figured out it could sell mobile phone on a slim margin, but then offer colored plastic faceplates as accessories that cost pennies to produce, but for which people would pay $20 in order to personalize their phones. The Finish wireless concern also whipped fellow phone makers Ericsson and Motorola due to its early realization that mobile phones would evolve into fashion items and that different users would want different features.

It's pretty ironic that Nokia's mobile phone operation was nearly sold. Although demand for cellular handsets was starting to take off, Nokia had little idea about how to raise production to meet the insatiable demand. It was Jorma Ollila, in charge of this division, who in 1991 sorted out production anomalies and found a way through the problems of getting the new GSM products ready for market. He quickly ramped up the assembly line and fixed the problem of components shortage. He had set eyes on the emerging GSM market since then. Ollila, a graduate from London School of Economics and a former Citibank executive, joined Nokia in 1985 and took over the reins at an almost bankrupt company later in 1992.

That year had seen the breakdown of Soviet Union, one of Nokia's major overseas markets, and the collapse of trade with the imploding Soviet Union had dealt a huge blow to Nokia. Shortly after taking charge of the company, Ollila made a strategic decision to focus business on wireless communications and began divesting the non-core operations. He gradually sold off Nokia's data communication, television and cable divisions, and started to pump up resources on GSM product development. With a global focus, Nokia built the R&D centers next to manufacturing facilities, a factor critical for an eighteen-month product life cycle of mobile handsets. More importantly, Ollila, a new broom, emphasized the significance of the company's own operational model.

Nokia's meteoric rise after the years of economic depression of the 1990s and the dot-com bubble of 2000 became a legendary success story. The Finish phone maker rapidly shifted from one handset model to another in order to fulfill new demands and sustained innovation in a cutthroat pricing environment by being closer to the customer compared to other wireless pioneers: Ericsson and Motorola. Furthermore, it strengthened its liaison with chip suppliers like Texas Instruments and cleverly tapped the resources of software companies in Silicon Valley. The timing was simply brilliant.

The Finnish electronics firm was spending all its energies on the cellular upstart GSM, and when it took off, Nokia was ready to meet the wireless operators' demands better than anyone else in the business. "When Nokia poured its resources into GSM, it was a moderately successful company from a small country betting against billions of dollars of entrenched infrastructure and widely accepted standards," *Wired* noted in 1999. "GSM took off—not only all over Europe but also in Asia, Latin America, and elsewhere." Ollila's focus for cellular pure play led to one of the most astounding success stories in the last decade of the twentieth century. On December 4, 1998, Nokia announced manufacturing 100 million mobile phones. Next year it passed Motorola to become the world's largest maker of mobile phones. Also, in December 1999, Nokia became Europe's most valuable company at the capital value of 203 billion Euros.

Nokia used to be a stodgy Finnish conglomerate, making everything from rubber boots and cables to lavatory papers and television sets. Even in that form, it had come a long way from its start in 1865 as a lumber mill on the bank of Emakoski River. But few would dispute the claim that Nokia had been Europe's outstanding business story. Politicians lined up to praise Nokia as an example of how Europe could prosper in the twenty-first century. Romano Prodi, president of the European Commission, drew attention to the success of Nokia and its Nordic neighbor Ericsson in a speech in 2002. "Their achievement in mobile telephones helped to create two vibrant clusters, around Oulu in Finland and Stockholm in Sweden, which have attracted a large number of startups as well as investment from foreign companies."

The transformation of this Victorian-era industrial conglomerate into a wireless powerhouse had become a Finnish fable. Fast forward to 2010 and the Nokia fairy tale had come down to earth. The next big disruption—smartphone—turned the Nokia brand of mobile excellence into an epic failure. The Espoo, Finland, company was suddenly at the risk of becoming a has-been, and watching Nokia slowly fall from the leadership role in the mobile industry seemed like carnage. What happened to one of the most celebrated corporate champions from tiny Finland? According to Henry Blodget—former research analyst and founder of news blog *Business Insider*—the iPhone happened. Smartphone was apparently the major cause of Nokia's troubles.

Apple redefined the smartphone market in 2007, and a few years later, Nokia found itself caught flatfooted. The European wireless stalwart had built its fortune on basic feature phones, but when the tides turned, even the most cost-conscious consumers were looking to buy smartphones. Poor leadership and complacency bred of success and compounded by an over-consensual culture caused Nokia to miss the smartphone revolution. Pride can kill a company and so can bad management; Nokia suffered from both and with a terrible timing. Nokia, once the emperor of mobile phones, shipping more than 100 million handsets per quarter, was in a tailspin in 2012. More painful was the fact that this paragon of mobile excellence was now feared to be the next Kodak.

The tech business history was littered with examples of companies that failed to deal with technology disruptions. Transitions had been a staple of the electronics industry, and in a way, Nokia's story, though fresh, was only a repeat of an old adventure. Kodak's woes, however, showed that technology trends—like shift to digital photography—were often clearly visible, but changing the course for a large corporate entity was exceedingly hard. Just like Kodak's internal teams argued for the digital shift that the top guys ignored, as explained in the fourth chapter, Nokia's senior management didn't heed engineers' outcry for Internet-centric, touchscreen-based app phones. It just took Apple and Google to show Nokia how to re-imagine the mobile phone with the help of powerful apps built on top of a slick computer-like user interface.

Nokia under the new leadership of Stephen Elop began frantically cutting costs and downsizing the company in a desperate bid to survive. The worst part about the announcements coming from Espoo, Finland was not the number of layoffs or the closed plants or refocused business units, but the fact that the industry had heard it all before. Kodak too had been in restructuring mode for fifteen years—cutting headcount and closing production and research facilities—before the digital imaging pioneer announced bankruptcy in 2012. Some Nokia proponents argued that the former mobile phone pacesetter still had a strong brand, a global presence and market share, but so did Kodak. Kodak's fate was a testament that no company, however strong, could count on continued success.

Xerox was another company that the industry observers began comparing Nokia with during the post-iPhone arena. Xerox, once an icon of the cutting-edge technology, developed many of the crucial building blocks of modern computing, including graphical user interface, Ethernet, and the modern text editor. Yet it was Apple who transformed the idea of graphical user interface into a commercially viable personal computer and changed the information technology landscape forever. Xerox continued to sell photocopiers while Apple stepped up the technology ladder to sell futuristic electronic appliances. Worse, the Norwalk, Connecticut–based company that once owned the photocopier industry eventually became a small player in what was essentially its core competency. Over the years, Xerox simply couldn't find traction to compete in a world with so many rivals vying for imaging products and services.

The analogies with the falling tech giants went on. Some pundits called Nokia the AT&T of the wireless arena. AT&T used to be the telecommunications colossus which controlled how Americans communicated with each other. The giant wireline phone company failed to adapt to the new market realities when the traditional wireline world gave way to the wireless and Internet playing fields. Some industry watchers even quoted Compaq and Digital Equipment Corp. as case studies to prove that it was hard for a company to stage a comeback once it became a dinosaur. But there were exceptions: Steve Jobs's Apple and Lou Gerstner's IBM. Gerstner arrived at IBM in 1993, slashed more than 100,000 jobs, killed uncompetitive OS/2 PC operation system, and began a massive shift from hardware to consulting services. Jobs returned to Apple in 1997, streamlined products, replaced most of the board, and expanded into new markets like music players and mobile phones.

Could Nokia reinvent itself like Apple and IBM? Or was it too late for the Finnish mobile company? It was one of those perennial questions that intrigued many in the industry. Nokia's smartphone travails showed a dismal reversal of fortunes where people stopped buying mobile phones and started buying mobile computers. Once Nokia saw itself being disrupted, the company brought in the first non-Finn chief in its 145-year history. A major upheaval was needed for this giant on the ropes, and one thing was clear: Nokia didn't have the luxury of time. Finland's stumbling mobile-phone titan seemed to be living on borrowed time, and the clock was ticking.

The aim of this book is twofold. First, it attempts to find out how Nokia—which was once so important to the smartphone business and wireless industry at large—reached at this crossroads. The book chronicles Nokia's lost decade in which the venerable handset champion spent nearly US$40 billion on research and ultimately found itself in the clutches of a vicious cycle. The book delves into one strategic blunder after another to provide a detailed account of this tale of management indecision. It chronicles how this comedy of errors took the Finnish mobile phone icon to a near-death experience. Second, could there be a triumphant comeback after this existential crisis? That is the recurrent question that the book tries to answer after having a microscopic look at the seemingly shaken Nokia's product and strategy roadmaps.

Nokia had a history of successfully adapting to big market shifts. The one-time wood pulp company had effectively diversified into electronics during the twentieth century, and it was in Nokia's DNA to willingly self-disrupt whenever necessary. The transition years of 2011 to 2014 marked a defining moment for the humbling Finnish corporate icon that was hardly a technological laggard. At this very crossroads, however, the Nokia story was engulfed in a plethora of misconceptions. A lot of information about the flailing handset maker was cluttered, and a number of facts were not in place. The book aspires to clear the air, develop a comprehensible picture, and thus set the record straight. Nokia was no more the master of the mobile game, but it was still an important company.

It all seemed so simple when Nokia's woes were seen through the eyes of skeptical analysts and sometimes overzealous trade media. But the reality check was a bit more intricate. The book digs deep into Nokia's heritage, strategy blunders, major stumbling blocks, and bailout efforts. That way, it attempts to recollect notes from this epic moment in Nokia's life and create an authentic document that not only recounts Nokia's breathtaking transformation, but also provides a discourse on the Finnish company's turnaround attempts. Here we go!

1 NOKIA COMES DOWN TO EARTH

"I believe we have lacked accountability and leadership to align and direct the company through these disruptive times. We had a series of misses. We haven't been delivering innovation fast enough. We're not collaborating internally."

— An excerpt from Stephen Elop's famous "burning platform" memo which he wrote to employees six months after he arrived at Nokia

In February 2011, Nokia's newish CEO Stephen Elop sent his staff an apocalyptical memo in which he likened the company to a worker standing on a burning oil rig who might jump into the icy water to escape the flames. A few days later, when Nokia announced a smartphone alliance with Microsoft, the Finnish firm just looked like that: a desperate man jumping into the icy water. A leading mobile phone player in terms of market share had embraced a mobile platform with 3 percent of the smartphone market. Nokia had effectively killed the Symbian—which, despite all its woes, still boasted the largest base of smartphone users—by turning it into a franchise platform and adopting Windows Phone 7 which was barely out of the gate. What agonized many in the industry was also the fact that, at the outset, it looked more like a business decision than a strategic one. Both Google and Microsoft made offers to Nokia, and in the end, Microsoft outbid Google in terms of engineering assistance and revenue sharing on mobile ads, search, and mapping services.

But adopting Android could have turned Nokia into another me-too handset house. Moreover, Nokia's biggest remaining software product—Navteq, a mapping service for which Nokia had paid US$8.1 billion—would have been brushed under the rug by Google, who had its own popular mapping service. Microsoft would integrate Navteq into Windows Phone software while its Bing search engine would power services in all of Nokia's devices. Nokia's content and applications store—Ovi— would integrate with Microsoft's Marketplace to provide more options to mobile users. It was a bold move for Microsoft in the sense that Nokia's economy of scale in handset business could solve Windows Phone 7's small market-share problem in one swoop if things went according to plan. But for Nokia, it felt like giving up, not like fighting back. It could even prove hugely damaging to Nokia's internal psyche to give up on its own software and turn to one of its competitor's.

The market sentiment had been overly pessimistic, and Google's vice president of engineering Vic Gundotra channeled that perception by saying "two turkeys do not make an eagle." The transition would take around two years, and during this time, Nokia expected to sell 150 million Symbian devices. But who was going to buy a smartphone with an obsolete operating system which, according to the company head, was not competitive enough. The partnership would take time to implement and deliver phones, and that could prove Nokia's Achilles' heel.

Windows Phone 7 was no one's priority until the world's largest handset maker committed to the platform. That made the logic of killing off Symbian to adopt a competing product with a lower market share highly defective. The critics argued that combining two companies' operations with flawed strategies wouldn't necessarily translate into shoring up each other's failings. Moreover, Windows Phone 7 had strict hardware requirements and allowed for little in the way of hardware design innovation, a core area of Nokia's strength. Nokia had also made a huge commitment to its developer community and had sunk millions into it. Its Ovi services center wasn't a smash hit like Apple's App Store, but it was making steady progress. And finally, though Nokia took a beating in the U.S. smartphone business, the opponents of the alliance argued that the American market wasn't everything. Nokia had huge brand awareness around the rest of the planet.

The work that Microsoft and Nokia had cut out for them was going to be long and tenuous, and all the while, the smartphone market was moving forward at a relentless speed. Nobody expected this transition to be a walk in the park. In 2011, Nokia made a pre-tax loss of 1 billion Euros. Then, a year after this landmark deal, Samsung displaced Nokia as the top mobile phone maker. And the bad news kept coming: the Finnish wireless firm had lost 90 percent of its market value since the launch of iPhone in 2007. The once-invincible wireless industry icon was undergoing one of the most spectacular crashes in the history of business. Dominating the smartphone market suddenly looked a faraway dream for Nokia; it was the company survival that was now at stake.

Elop's disruption memo was a lengthy and ultimately a public declaration that dissected the threats to Nokia's competitive position. Next, the new CEO spotted how long the platform had been burning and decided to go for a radical strategy to start over with a modern mobile operating system. There was quite a bit of enigma in the trade press on what happened to this undisputed heavyweight of the mobile phone world. To find answers let's first go back to 1996 when Nokia was on top of the world and its smartphone dream was just beginning to take shape.

HERE COMES THE BEAST

Star Trek, first aired in 1966, fascinated the world with its sophisticated collection of gadgetry that included communicators, phasers, and tricorders. The iconic television serial drew inspiration from a century-old strand of technological utopianism prevalent in American science fiction. *Star Trek* was nothing without its optimism about technology. Much of the technology displayed in it was abundantly imaginative; pundits called it twenty-fourth century inventions. At that time, hardly anybody could have imagined that the world would be seeing such gadgets in real-life working environment just in a few decades.

In the late 1990s, at the height of cellular boom, tech industry innovators began conceiving *Star Trek*-like gadgets alongside the constraints of contemporary technologies. They were convinced that a vintage *Star Trek* communicator was, in fact, equivalent to a modern-day cellular handset. At the 1996 CeBIT fair, thirty years after the inception of *Star Trek*, mobile phone bellwether Nokia, which had won the handset wars by turning an expensive, gray business tool into a cheap, colorful consumer product, exemplified the nascent wireless data market by introducing the first multipurpose device on commercial scale. The Nokia 9000 Communicator was a large and pricey product that combined voice telephony with an organizer function and a modest Internet access.

Nokia's Communicator had identified the rise of a new segment in wireless data: the smartphone. This distinctive palmtop computer-style smartphone was the result of combining a modestly successful and expensive personal digital assistant (PDA) from Hewlett Packard with Nokia's bestselling phone. In fact, the early prototype model had these two particular devices fixed via a hinge. The Nokia 9000—a shotgun marriage of a GSM mobile and a palmtop PDA—had just 8MB of memory and weighed slightly under a pound. It looked like a regular phone from the front until it flipped open to reveal a second screen and a QWERTY keyboard. The handset-cum-organizer made use of an operating system (OS) developed by the California–based company Geoworks Corp. Nokia's Mikko Terho was a leading influence behind the development of arguably the world's first commercial smartphone.

Shortly after this launch, Nokia released an enhanced version with smaller size. The Nokia 9110 Communicator could download web pages during off-peak hours and connect to a digital camera for transmitting pictures. The glitzy device weighing 253 grams got a much-publicized product placement in Val Kilmer's 1997 movie *The Saint*. Although 9110 Communicator was an expensive product with a price tag of US$1,000, by enabling services like fax and e-mail on a hybrid platform, it accomplished an initial recognition for a new category that could gather momentum in the future. Earlier, to complement such services, Nokia had produced the first cellular data card and software-based data suite for GSM phones.

When the next sequel, Nokia Communicator 9210, arrived in the market on November 21, 2000, reviewers likened it to a 1980s' science fiction movie villain. Branded as the Communicator 9290 in North America, it was the first smartphone with an open operating system—Symbian. The Nokia 9210 was a combination of mobile phone, organizer, and e-mail and web device with a color screen and a keyboard. The champion in designing cool wireless handsets had essentially rebuilt a Psion handheld computer as a mobile phone that made some initial grounds in Europe. The Nokia 9210 opened like a book and looked more of a little laptop that the user dug out of a briefcase or large purse, rather than a phone kept close at hand. Beyond its bulk, it was also slow in PC synchronization and downloading e-mails.

The fact that the new data-centric gadgetry was to be based on Internet-style packet switching rather than circuit switching of conventional voice networks meant that even the least sophisticated handsets could be far more complex than high-end mobile phones of the 1990s. Not surprisingly, therefore, Nokia's reentry into the commercial smartphone market with the launch of US$600 Communicator 9210 was hailed as a weak start. The astronomically expensive price bracket alienated the majority of public. In retrospect, though not very successful, the unfolding of these devices started a transition from a voice-only strategy; until this point, the mobile phone industry had been built on a single application: voice. The goal was to create a smart gizmo that would allow people to check their e-mails, surf the Internet, plan their schedules, and, of course, make phone calls: in other words, an electronic organizer, a personal computer, and a mobile phone—a device that nobody could afford to be without.

The Nokia Communicators became a niche hit; the handsets drew a dedicated following among certain business users, but never commanded a mass audience. These early products fell short in one or more significant ways, and the consumer often bypassed them in favor of a separate cell phone and PDA. The Nokia Communicator models were also remarkable in that they were the most expensive phone sold by a major brand; sometimes 40 percent more pricey than the next most expensive smartphone by any major manufacturer. Interestingly, although the Nokia 9210 was arguably the first true smartphone with an open operating system, Nokia continued to refer to it and the following models as Communicators. Ericsson referred to its GS88 handset as "smartphone" for the first time.

SMARTPHONE HANGEOVER

In a country where people aim to get things done before boasting about it, Nokia, true to its Finnish roots, didn't create a marketing buzz while launching these high-tech toys, calling Communicators a mere new category that could evolve into a more meaningful product in the future. That was quite unusual in the telecom world known for its hyper-marketing practices. Such an unconventional, modest, but practical approach had been credited for leading Nokia to an unprecedented stardom in the digital wireless realm. Its ability to think about the future had so far been a prominent factor in its startling transformation. On the flip side, however, when the notion of the mobile Internet with software portfolios began to crystallize, instead of pushing cell phones as next-generation Internet appliances and inventing innovative new products, Nokia remained focused on shaping the entire handset market.

Back then, as later recalled by Jorma Ollila, the company had exactly the right view of what smartphone market was all about. He ran Nokia as CEO from 1992 to 2006 and had been the company's chairman from 1999 to 2012. Ollila claimed that the Finnish mobile handset maker was about five years ahead of the smartphone market. But ultimately the Nokia Communicator—a pioneering forerunner of the iPhone era smartphones—became a stark reminder of how Nokia failed to deliver on its great potential. The Communicator included the ability to download apps, but Nokia managers felt that downloading apps was something only a minority of people would do. A few years later, however, the world saw Apple doing a mainstream TV commercial about downloading apps onto mobile phone.

It's pretty ironic that before the iPhone had apps and Android got Maps, Nokia handsets had both. Amid this reversal of fortunes, some market watchers claimed that Nokia was an engineering company that needed more marketing savvy. But that wasn't quite true. Nokia had been acclaimed for its marketing, and was seen as the company which had figured out how to turn mobile phones into fashion accessories. Nokia had also been quick to pick up the Finnish design aesthetics, but its engineers were expert at building physical devices, not the software that made those devices work. The company's development process was largely dominated by hardware engineers and software experts were marginalized. To Nokia's credit, however, it anticipated shift toward software and services much earlier than other handset makers.

It launched Ovi store in 2007, almost a year before Apple opened its highly successful App Store. A few months later, Nokia bought Navteq, a maker of digital maps, to be able to harness the power of location-based services. Shortly thereafter, Nokia launched "Comes with Music" service using a handset pairing with a digital music subscription. But despite the fact that Nokia correctly predicted the evolution of the market toward software and Internet services-driven product differentiation as early as 1998, its execution had been disappointing, resulting in substantial high-end market share loss to Apple, and increasingly, to Android.

The key issue facing Nokia was how to tie together its hardware and software into a compelling package, as Apple had managed to do. Nokia was a hardware maker that wanted to be a software and services company, but its top management grossly underestimated the challenges related to moving from a hardware-driven business model to the software world. It simply took too long for Nokia's software and services strategy to unfold. Like Sony, Nokia couldn't find a way to shift from hardware to software business. In the final analysis, Nokia was a hardware-focused organization that never mastered its software strategy. Its software roadmap was in a continuous state of flux if not downright directionless.

Ollila admitted in an interview with the Finnish newspaper *Helsingin Sanomat* in October 2013 that the problem was recognized already in the 1990s and that there were plans to fix it, but they were not implemented. "The Mobile Phones Unit once had 1,000 people in Silicon Valley, with the task of picking up on new trends of software development. But we were not successful in the way that Google and Apple were later. That is the key failure for Nokia." He also recounted that there were assurances from different parts of the organization in 2008 and 2009 that Symbian could be upgraded and made competitive.

Moreover, Nokia seemed to have underestimated how important the shift to smartphones would be and that failure partly resulted in an institutional reluctance to transition into a new mobile era. Nokia also overestimated the strength of its brand and believed that it would be able to catch up quickly even if it was late to the smartphone game. The story of Nokia's fall was practically the story of the smartphone's rise. And the fact that the Finnish mobile firm remained in a denial about this tectonic shift for too long. Nokia was nonplussed when Apple launched its iconic iPhone device in 2007, partly because it had over 40 percent of the market share at that time.

During the late 1990s, telecom companies, software houses, and handset makers had all started revving up strategies for the next tech revolution: wireless access to the Internet. They said portable and constant link to the information world would change lifestyles—and offer business opportunities—just as much as the wired web. Apparently, for Nokia, the market leader, to bring handsets the advanced capabilities to support data services like the Internet access, the next step was to develop a powerful operating system for mobile phones. A mobile operating system was the underlying software that guided the most important functions of a smartphone. Mobile handset makers took the baseline operating system and built features into and on top of it.

A decade later, when Nokia's first big software venture took shape in the form of Symbian mobile operating system, most of its allies and competitors had joined the Android bandwagon, getting a head-start both in terms of cost and time-to-market. Now Nokia looked like Alice in Wonderland,

and its smartphone ambitions were stuck in a space that was marked with endless gamesmanship for winning the hearts and minds for its Symbian prodigy. Critics blamed Nokia's Symbian operating system for the company's failure to move ahead in smartphones, saying it was so clunky that developers had not been willing to write applications for it. The next couple of chapters are dedicated to Nokia's struggles in the mobile software labyrinth.

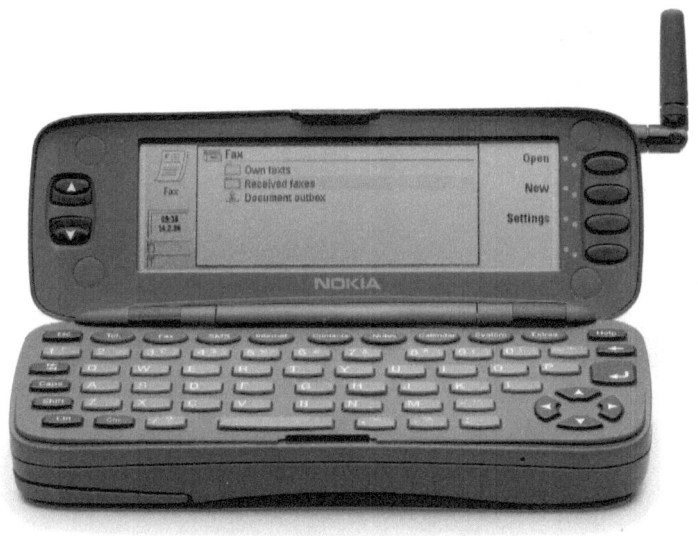

The Nokia 9000 Communicator, according to Psion founder David Potter, was a bird that couldn't fly.

Photo courtesy of Nokia

The Nokia 9110 Communicator got a much-publicized product placement in Val Kilmer's 1997 movie *The Saint*.

Photo credit: Nokia

When Jorma Ollila took over in January 1992, the company was teetering on the verge of a collapse. He steered the Finnish company to become the world's biggest maker of cell phones. He had also set his sight on ushering the world into the era of smartphones, but Ollila and his successors failed to deliver on that promise when the smartphone era arrived.

Source: www.iltalehti.fi

2 | THE SYMBIAN SAGA

"It was an ignorant complacency, not an arrogant complacency."

— Juha Äkräs, Nokia's head of
human resources

In June 1998, Nokia and its Scandinavian cousin Ericsson embarked on a venture with British handheld computer maker Psion PLC to develop a common software platform enabling mobile phones access the Internet and other data services. Motorola joined the consortium—Symbian—at the last minute. Symbian announced a schedule to release the first OS-based mobile phone in a three-year time frame. The claim that its still unfinished operating system would be the new standard for smartphones and wireless Internet appliances created a stir in Europe and across the world. The alliance had got a head-start with support from all three wireless industry titans.

When Apple's charismatic chieftain Steve Jobs launched Macintosh in 1984 with the mantra of "The Computer for the rest of us," a tiny British startup came with its own slogan: Computer in every pocket. Back then, hardly anybody took notice of Psion, the company founded in 1980 by a Zimbabwean-born electrical engineer David Potter. Psion's Series 3 handheld computer had scored a hit where much richer U.S. companies had failed. Psion—a publically traded firm at the Edgware Road in London—was unanimous in its conviction that company's future belonged to mobile data communications. The pioneering British outfit strove for the quality and conviction borne out of Europe's expertise in Global System for Mobile Communications (GSM) standard.

A small but highly focused firm, Psion started with manufacturing handheld computing products. Later, in the mid-1990s, it began developing an organizer operating system it called EPOC. In November 1994, Colly Myers, along with two engineers, had started working on an operating system that could match the capabilities of the mainframes but also cope with the extreme demands of low-power mobile computing. There was this realization that the long-term prospects of Psion Series 3 successors hinged on company's own software. At the same time, however, this realization couldn't justify the investment in full-fledged operating system software for just one customer: Psion. So the company was split into two entities, and the OS software outfit began to pitch the unreleased operating system to prospect licensees.

Nokia liked the software under development and decided to use it for the upcoming Communicator instead of Geoworks'

GEOS operating system software. When Psion approached Nokia managers for licensing Psion OS software, they were apparently excited and said, "Let's go to Ericsson." They also wanted to get Motorola on-board but excluded the Schaumberg, Illinois–based firm from early negotiations amid fears that its bureaucracy could derail the talks. There were competing teams within Motorola's mobile operations, and each team had a choice to pick its own operating system. Later on, after the basic Symbian blueprint had been hatched out, Motorola was given forty-eight hours to sign up. "We said, put your name here, on the same basis as Nokia and Ericsson. Nobody will ever know you weren't here from the start. They signed it, but hated the fact they were dragged into it," recalled Stephen Randall, a key Psion figure in the making of Symbian.

The notion of the commercial smartphone erupted on the wireless scene in 1998 only when Symbian promised to transform the whole idea of mobile gadgetry through phones with PC-like software programs and e-mail functions. The Symbian venture was to create a standard built around Psion's EPOC operating system for wireless information devices such as smartphones and PDAs. Psion quickly established itself as the best horse to back as its EPOC operating system was seen as the most impressive technology for the mobile Internet age. The EPOC software was perceived as especially suitable for mobile devices because it claimed to make better use of battery power and memory, a crucial advantage in the wake of high usage of these resources in non-voice communications.

Phones were to become mobile computing platforms with a collection of operating systems, application suites,

browsers, and user interfaces. So more and more importance now resided in software. That made Symbian one of
the most important technology companies on the planet.
Symbian remained a darling of the technology press for the
first two years of its life. But Symbian by no means marked
the beginning of the smartphone era.

In spring 1995, a year before the launch of Palm Pilot,
Jeff Hawkins had visited Ericsson's headquarters outside
Stockholm to present his views on a mobile phone that
could evolve into a smartphone. The Ericsson managers
were intrigued. Ultimately, when the stakes for the brain of
next-generation mobile devices got high, so did the efforts
of the PDA pioneer Palm. The popularity of the Palm operating system made its proponents believe that the software
for portable computers could be quickly adapted to bring
wireless capabilities. And Palm, like Symbian, was based on
an open standard. Moreover, Psion's keyboard-based handheld devices had been losing market share to Palm-like
pen-based systems.

Then there was Nokia's sibling rivalry with Microsoft, which
later became its most committed partner in a strange
about-face. At the height of the Internet boom, in July 1999,
luminaries of technology and media businesses gathered
at Sun Valley, Idaho for a high-tech retreat. Over there,
according to a story published in the July 9, 2001 issue of
BusinessWeek, Microsoft boss, Bill Gates, approached Nokia
chief, Jorma Ollila, to discuss the mobile Internet and how
their respective companies—one a titan in software, the
other in mobile phones—were preparing for it. Gates was
hoping that the Finnish handset maker could incorporate

Microsoft software in its phones. "How come we don't merge our efforts with Nokia?" he later wrote to his colleagues.

But Ollila and his comrades had seen long ago how Microsoft virtually took over the personal computer industry; so even when Ollila and Gates chatted, the Finns were moving to keep the PC behemoth from Redmond at bay. What probably motivated Symbian's parents more than anything else was that as computing and wireless converged, they didn't wish to end up like PC makers—low-margin assemblers that were little more than a distribution channel for Microsoft's intellectual property. On the other hand, Microsoft, seeing the PC's reign at the top of the electronics food chain falling apart, wanted to thrive or at least survive in the post-PC era. Bill Gates had singled out the Symbian consortium as one of the greatest threats to his company.

TROUBLE IN PARADISE

Microsoft was keen to ensure a key role for itself in the embryonic smartphone market but its hypercompetitive market drive was apparently a non-starter for the wireless industry. Microsoft, however, was not easily thwarted. It started seeking handset partners and licensees, and joined hands with wireless operators such as British Telecom and NTT DoCoMo. The PC software giant was doing everything in its powers to ensure that Windows CE—a slimmed-down version of the operating system that ran on personal computers—would find its way into mobile phones.

But while Symbian was tightly focused on the wireless market, Microsoft wanted to turn its Windows CE software into an operating system for a whole range of electronic devices. The result, according to some software experts, was an operating system that was too clumsy for mobile phones. A friend of all is a friend of none! Rivals called Windows CE merely a small computer platform. Then there was the stance that a lot of communication software was involved in a phone and that Microsoft was not really a communications company.

Smartphone was an inflection point that presented starkly different approaches from telecom and computer bellwethers. In the early phase of the smartphone inception, the computer industry tried to cram PCs into pocket-sized devices, while the mobile phone industry had arrived at the same point by adding data-centric features to handsets. When the once distinct worlds of computing and mobile telephony began colliding and converging, the giants of each industry—Microsoft and Nokia, respectively—squared up for pre-eminence.

In those early years of the smartphone's awkward adolescence, the most likely outcome was conceived as a fragmented market in which Symbian would hold on to smartphones and other phone-like communications devices hooked to the Internet. Palm would continue serving the mass market for handheld devices, and Microsoft would focus on more demanding business users. That scenario was also seen as leading to a parting of ways between Europe and the United States. The early smartphone models had made modest inroads in Europe and Asia, while America, where lack of a single digital wireless standard hampered the growth, initially looked stuck to phoneless organizers by using them for data-only or pager-like connections.

In all fairness to Nokia, it gave Symbian the strategic worth that the smartphone initiative deserved, but in the end, the overall execution stumbled. Few realized early on that Symbian was going to be a product designed by a committee of archrivals. Symbian had another major flaw in its anatomy: it had two levels of management. While the operational board ran day-to-day affairs, owners made decisions behind closed door and then enforced them on high. Fed up with Symbian's bureaucratic structure, several key staff members quit the joint venture, and Symbian's work on new phone software started slipping behind schedule.

Symbian's plans had stalled a couple of years after its inception. Product delays and lack of commercial focus had beset Europe's bright hope in portable data communications. Symbian began to see the U.S. rivals eating away its promising early lead. In the early going, EPOC operating system drove fewer than 5 percent of the handheld computers and smartphones sold worldwide. Now the big three mobile phone makers—Ericsson, Motorola, and Nokia—were looking in other directions, and as a result, Symbian was no longer the front-runner in the operating system race for the soul of next-generation mobile handsets. Fast forward to 2010, trade media was playing the Apple versus Google riff much in the same way they sang the siren songs of an imminent battle between Nokia and Microsoft a decade ago.

NEW MOBILE ORDER

In retrospect, while Nokia and Microsoft fiddled with their software strategies during the mid-2000s, Apple and Google were quietly making strides with iOS and Android operating systems, respectively. Apple and Google showed users what an OS platform could do for mobile phones, a premise where Nokia and Microsoft clearly struggled. Both Nokia and Microsoft had been in this space since the beginning. It was theirs to lose, and they lost it. They had everything to create the right environment for powerful applications and a compelling user experience. Symbian and Microsoft could be credited with first mover status in the smartphone industry, yet despite their valiant efforts to establish third-party development programs, neither could achieve the kind of success that Apple and Google accomplished with their respective software platforms.

Apple, instead of desperately trying to defend comparisons of the iPhone platform to Windows Mobile in features or Symbian in reach, focused on ease of use and the unique value it was adding. And rather than ceding control of the App Store to third-party developers, Apple maintained influence over its development environment to avoid allowing the software platform to develop a reputation of being crass, seedy, sloppy, or unfinished. "We weren't the first to this party, but we're going to be the best," Steve Jobs told his audience at the launch of iPhone OS 4.0 software in April 2010. Apple stoked the smartphone market with its iOS platform and rallied developers behind this solid operating system based on a subset of its well-known desktop software.

It'd be worth mentioning here that Apple—the computer maker that had little to do with mobile communications—developed the software first and phone later. Apple's Benjamin Button approach for its iPhone operating system turned out to be a stroke of genius. The iPhone OS got the single most important thing—user interface—right from the start. The user interface is the presentation layer of the OS software. In the hindsight, looking back at the agony of Symbian, which had been there since the early 2000s, running on millions of phones, software experts were almost unanimous in their belief that if there was one fundamental problem with Symbian, it was the user interface. Otherwise, they said, Symbian was still the best OS from a technical standpoint as it was robust and consumed very little power.

It became apparent when Google's first Android-based phone—the HTC G1—came in 2008 offering a user experience similar to the iPhone. Next up was RIM—which later renamed itself as BlackBerry, its best-known product—torn between enterprise and consumer markets; its engineers burned the midnight oil to furnish touchscreen interface on the new smartphone models. Even Microsoft dumped Windows Mobile altogether and reinvented Windows Phone 7 operating system to have a similar user interface with new software chops.

Symbian didn't deliver the web experience or trendy apps that Apple had already going for it. But Google was another matter. The twenty-first-century poster child of Silicon Valley innovation had managed to create a platform that genuinely rivaled iPhone for the best smartphone experience around. Initially, Apple seemed to be singlehandedly winning the smartphone battle, but developers eventually acknowledged that Google's Android would shape into the second best bet. In retrospect, the iPhone was just a harbinger of Nokia's smartphone problems. The Finnish company's real undoing was the rise of Android that provided Nokia's mobile phone rivals a proxy platform and reclaimed the leadership in mobile handsets.

Android emerged on the mobile scene in 2008, and just in two years, it took the leadership position in smartphone volumes. By 2012, according to IDC figures, Android was on an average of three out of four smartphones sold worldwide. When Android took the mobile world by storm, Nokia didn't have a single product to battle Android original equipment manufacturers (OEMs). The emissaries of Android army—HTC, Motorola and Samsung—were all outselling Nokia in the smartphone market. The grim details showed that Android's rapid rise had mostly come at the expense of Nokia; it was Android that mostly absorbed Nokia's losses around the world.

The smartphone market had clearly gravitated toward platforms on top of which third-party developers could build apps that ran on mobile handsets. If history is any guide, platform markets tend to standardize around one or two winners, which at the time of smartphone industry's takeoff, were Apple's iOS and Google's Android. Predictably, apps marked the next chapter of Nokia's software miseries. In a perfect irony, more than a year before the launch of Apple's App Store, Nokia had introduced Ovi as a bundle of Internet services that let mobile users keep calendars, access files, download music, back up contacts, and carry out other activities.

The store featured games, applications, podcasts, and videos for smartphones running Nokia's Symbian operating system. Still, the Finnish mobile titan found itself flat-footed in the new apps economy. Nokia's Ovi Store for mobile apps got off to a rocky start as users faced problems accessing the store and downloading the programs. The Ovi interface was challenging to navigate and its cluttered offering of services like "Comes with Music" was confusing. Nokia opened the app store that was geared to more technically savvy users, and was not as easy to use as the Apple App Store. Moreover, the Finnish mobile powerhouse didn't put sufficient marketing muscle behind it.

Ovi—which meant "door" in Finnish—was the critical linchpin in Nokia's quest to wrestle with new-age smartphone rivals like Apple and Google. But Ovi found itself in a vicious cycle soon after its re-launch in 2009. The apps phenomenon mostly related to the U.S. smartphone market while much of the support for Symbian came from European developers. The developers in the United States rushed to write programs for the iPhone and Android, but shunned Symbian. With Nokia's low market share in the U.S. smartphone business, there was no point for a developer to invest time and money to create an app for the Ovi Store. Creating apps for different platforms often requires learning a new programming language. Moreover, to make a return on investment, developers focus on the platforms that provide the most access to users and revenue. Apple's iOS and Google's Android were evidently fitting the bill.

Apple, in a stark contrast to Nokia, carefully rationalized the supporting technology components, worked out a robust product roadmap, and then mobilized its legendary marketing machine. It's official now that the iPhone was initially conceived as a phone that would play music and video from iTunes, but its primary appeal quickly became the millions of head-slapping, useful software applications that ran on it. The iPhone certainly played music, but owners were just as likely to use it to check the weather, book dinner reservations, read a newspaper, get directions, or play a quick game of Taxiball on the subway. The creation of the App Store for the iPhone was a revolution that literally changed the smartphone world overnight and helped the iPhone reach dizzying heights.

SYMBIAN AT CROSSROADS

The anecdotal evidence presented in the next couple of chapters affirms Symbian's cumbersome legacy software and Nokia's predominantly hardware-centric organizational model. When Android and iPhone were taking over the gadget mindshare in the late 2000s, even pared-down models needed eight times as much memory and processing power in Symbian handsets. Apple and Google were relentlessly eroding Nokia's smartphone market share by bringing cool new stuff to mobile phone users. For software guys looking to develop apps for iPhone or Android, all they needed was download the software developers kit (SDK), register, pop open Xcode (for iPhone) or Eclipse (for Android), and they could get started.

But if they wanted to create an app for Symbian, there were just too many choices: they could develop in native Symbian, in some Java variants, in .Net integrated, or in Qt.

Nokia had so many versions of Symbian and development environments that apps created for one set of devices often didn't work on another. And that led to fragmentation similar to a degree to what industry later witnessed in the case of Android, though for a different set of reasons.

It became far more challenging to develop an app for Symbian compared to Android, let alone the iPhone. Take the example of an app that turned consumer photos into postcards. Developing such an app for Symbian could be a long road because the developer would have to build a plug-in that connected the camera to the gallery of photos; the app could take about four to five weeks of work. On the other hand, in Android environment, it would take just five minutes. It was a feature built into the operating system, and all a developer had to do was turn it on.

Technology waits for no one. The market was moving quickly, and Nokia urgently needed to deliver an exciting and genuinely differentiated device to regain its reputation as an innovative technology leader. So far, Nokia's excellent distribution and sales strategies had ensured market share in the mobile OS domain, but the company understood too well that status quo was not a given in these quicksand times. Nokia's 2010 shift to feature MeeGo operating system software in its upcoming premium phones was a clear sign of growing strains and marked a gradual shift away from Symbian.

Symbian had been a victim of first birth and great expectations. Its dominance faded as mobile software platform market became more crowded. When Nokia started calling up developers on the reinvigorated Symbian Foundation platform, there weren't many listeners. Despite the fact that Nokia was still the top smartphone maker in terms of market share, developers were flocking to competing platforms— Android and iPhone operating systems—as they saw a negative momentum with Nokia. These software types have a very keen insight of how fast or slow a platform is gaining traction. The way Android had knocked down royalty-based models of Windows Mobile and Palm, and had forced the 800-pound gorilla Symbian to toe its "open-source" line hadn't gone unnoticed. Apple was a different affair though.

Symbian became hopelessly buggy as Nokia tried to make the ten-year-old dog pull off iPhone-like tricks. Nokia, at a time when mid-range feature phone market was the sweet spot, saw Symbian mostly as a feature phone platform with a little bit extra. However, when the company realized it needed to do more, Symbian was a bit too old and wasn't extendable enough to do the things that Nokia needed to do. With younger and energetic competitors in the market, Symbian seemed like an aging actress that should have already stepped out of the spotlight. It continued to decline until Stephen Elop terminated the Symbian program by hooking up with Microsoft.

It was a clear admission that Nokia's mobile platform strategy had faltered. The poor execution reflected in a number of factors, including an endless gamesmanship for the Symbian revival and a lack of services expertise. Symbian was all about software and Nokia messed up the project by not being timely in its code releases. The Espoo, Finland–based firm also demonstrated a lack of attitude toward compatibility of apps and broke backward compatibility on OS upgrades on multiple occasions.

This chapter attempted to chronicle Nokia's broader journey into the smartphone realm and find out why Nokia—which correctly predicted the evolution of the market toward software and Internet services-driven product differentiation as early as 1998—ended up as an also-ran. Most of Nokia's problems could now be traced back to Symbian. But why Nokia fought a losing battle? How did Nokia reach a dead end? To find all these answers, it'd be worthwhile to have a closer look at the Symbian story. The next chapter will document Symbian's slow death in the wake of more modern user interfaces offered by iPhone and Android devices. It's an epic tale of disruption that started with the birth of an industry that Nokia sowed with its own hands.

MikroMikko—a personal computer produced by Nokia Data division from 1981 to 1987—was the Finnish company's attempt to enter the computer business. In 1984, Bill Gates visited Nokia to check out the MikroMikko desktop computer. However, in 1991, Nokia sold its personal computer division to the British firm International Computers Ltd (ICL), which later became part of Fujitsu. No one could have imagined back then that one day Nokia would desperately need its computing DNA to deal with mobile computers a.k.a. smartphones.

Image Source: Nokia Data

Symbian, which affected Nokia's future in a strange way, was the prodigy of Psion software. A mid-1980s photo of Psion's software team shows younger Charles Davies (second from left) and Colly Myers (third from left). Davies, Psion's employee number one, later became Symbian's chief technologist. Myers, the designer-in-chief of EPOC software, had been Symbian's CEO from its foundation in 1998 to 2002.

FROM SYMBIAN TO MEEGO

"More and more, we are becoming a software company."

—Jorma Ollila, Nokia's chairman and chief
executive officer in a 2001 interview

Symbian's history is a long one. As Andrew Orlowski narrated in his article "Symbian, the Secret History" published in *The Register*, the first major turning point in the Symbian story came with the decision to offer four user interfaces (UIs) that could cover the whole array of mobile devices. The planned user interface design count eventually came down to three: Crystal, Pearl, and Quartz.

The Crystal design related to QWERTY keyboards and shared the Nokia Communicators' look and feel; the Pearl design covered all one-handed soft key designs; and the Quartz was a pen-based user interface which had its origin in Ericsson's design lab in Ronneby, Sweden. In retrospect, that only confused the nascent smartphone market. In the meantime, Nokia, then the smartest kid on the block, came to this realization that if it could own the UI part, it could own the user experience, and ultimately the developers. And that's where the value was going to migrate, reckoned Nokia managers. So Nokia, executing flawlessly in those days, subsequently led the Symbian partners into this game-changing decision that the venture had to be limited to only developing base code for the mobile phone operating system.

For Symbian alliance, that meant abandoning the work to develop application software, which provided the underlying features essential to a smartphone operating system, such as support for graphics, security, and Internet access. Through that application software or user interface, licensees could also change the phones' on-screen menus and other features like graphics. Another interesting twist in the Symbian archetype emerged when Nokia brought the Pearl portion of the project in-house and replaced it with its own software—later dubbed as the Series 60—to control such applications as picture messaging, web browsing, and e-mail. Nokia took the softkey UI from Symbian and eventually converted it to keyboard-only software.

In the article "Facing Big Threat From Microsoft, Nokia Places a Bet," published on May 23, 2002, *The Wall Street Journal* chronicled how Nokia's initial plan was to keep the Series 60 software to itself. The strategy could preserve its profit margins but at a potentially high cost. The move could make it harder for Nokia to stay ahead of Microsoft in the race to establish a mobile software standard. Apple Computer had taken a similar approach with its PC software by keeping it proprietary and wound up as a niche supplier once Microsoft's Windows took off. So Nokia chose the U.S. computer show Comdex in November 2001 to announce that the Series 60 would be available to all comers. The top mobile phone maker was at pains to persuade the rest of the industry that Nokia wasn't secretly plotting to become a monopoly. Senior company executives pledged that Nokia would license the software at very low rates and make the source code—which showed exactly how Series 60 was written—available to licensees.

Nokia, for its part, was looking to cement its Series 60 software platform on top of the Symbian operating system as a de facto standard for smartphone applications. The company added an abstraction layer to the Series 60 by using web runtime tools and Qt development platform so that the Series 60 software—employed by application developers—could also run on other mobile operating systems. Qt, pronounced "cute," was an open-source application platform for creating software that worked on multiple platforms. "We need to agree on a common architecture for the middleware on top of the operating system to make next-generation services and applications happen," Nokia chairman and chief executive officer Jorma Ollila told the Comdex crowd in Las Vegas.

In the coming months, the company licensed the platform to Siemens, Matsushita, Samsung, and Sendo. When Nokia added itself to the list, the companies that signed to the Series 60 platform collectively controlled more than 60 percent of the handset market. Nokia made a great play on the fact that the Series 60 was significantly more OEM-friendly and that it allowed handset makers to customize their offerings. In the meantime, Symbian OS software, the base system that helped Nokia succeed with its Series 60 platform, was ready for the next big shake-up.

In 2004, Nokia took control of Symbian through the purchase of Psion's 31.1 percent stake valued at an estimated US$252 million. At this stage, Ericsson CEO Carl-Henric Svanberg urged minority shareholders in Symbian Ltd to use their proportional preemption rights to stop Nokia from gaining a majority stake in the mobile phone OS supplier. But Nokia went ahead anyway. The future of the Symbian group had been further clouded after the largest shareholder in Psion announced its opposition to the sale of its Symbian stake to Nokia. Earlier in late 2003, Nokia had bought out Motorola's 19 percent stake, which clearly suggested that it was preparing to take a controlling stake. The quest for the control of Symbian was seen as a move by Nokia to strengthen its position ahead of a looming battle with Microsoft in the use of operating systems for next-generation smartphones.

It was a time of a dramatic realignment given that the telecom bubble had burst and limitless money that the Symbian alliance promised had dried up. Nokia's Nordic cousin Ericsson was experiencing a staggering fall and had spun off its mobile phone business into a partnership with Sony in 2001. Motorola had also fallen into a funk. Then there were WAP and 3G fiascos lying at the wireless industry doorstep. It was precisely this set of circumstances that helped Nokia—still riding high on a strong volume of handset shipments—to become so good at getting what it wanted from the Symbian alliance.

There was apparently a merit to Nokia's hardball approach. The standardization by consensus, the hallmark of the old Symbian, was seen as giving proprietary systems such as Windows Mobile and Palm an advantage in time-to-market and nimbleness. On the other hand, the move had the opposite effect as many licensees of Symbian began to perceive the operating system as Nokia's proprietary technology rather than an open standard. An analyst called the move "the ultimate manifestation of the boy scout effect." So far, Symbian had been a tightly controlled ecosystem where fragmentation had not been allowed to happen. But this control point was now in danger of being breached.

By 2006, nearly 100 million Symbian phones were in the market, a figure that more than doubled two years later as the smartphone market continued to grow. Until this time, Symbian remained an undisputed market leader in the smartphone space, largely owing to the strength of Nokia. But Symbian saw the mobile landscape changing once the iPhone started to rewrite the rules of the wireless game. Charging a royalty for use of its software and requiring its partners to sign up to a license agreement eventually became a barrier, which for a partner was a financial burden as much as an operational overhead. Google was clearly aiming at that stumbling block through its free operating system software: Android. Consequently, the next crucial stage in the evolution of Symbian came in 2008, when Nokia purchased all Symbian assets and started to guide the software down to the path to open source.

In summer 2008, Symbian co-founder Nokia announced that it was buying the 52 percent of the software maker that it didn't already own and would release the mobile operating system under an open-source license. To support the new open-source project, Nokia established the Symbian Foundation, a collection of hardware and software companies that pledged to donate code and resources to Symbian's development. Phone makers Motorola and Sony Ericsson got on-board, contributing software from their UIQ project, which had evolved from the pen-based Quartz interface for Symbian. The Japanese carrier NTT DoCoMo promised support while contributing its MOAP (S) interface. Other supporters included AT&T, Samsung, STMicroelectronics, and Texas Instruments.

SOFTWARE CONUNDRUM

Nokia's acquisition of Symbian was initially hailed by some industry corners as a game-changing move. They advocated that by turning Symbian into a non-profit entity—Symbian Foundation—Nokia had ushered into its future the collective power of the open-source community. By making Symbian an open-source operating system, Nokia engineered a brilliant preemptive strike against Android, or so they said. A closer look at the making of Symbian Foundation revealed a different story. The Finnish mobile phone giant, by pursuing a two-pronged strategy, was seen by some industry sections as trying to have it both ways. It was championing an open philosophy while fostering software unique to Nokia.

The very problem for Symbian's future was that its success was intricately tied to the success of Nokia. In other words, the dilemma was that while it was inevitable for Symbian Foundation to go beyond Nokia, at the same time, Symbian looked bound up with the fate of Nokia. If Symbian had historically been bound to the fate of Nokia, part of the unspoken rationale behind the creation of the Symbian Foundation was to go beyond Nokia. Even if, in the short term, this meant getting closer to its former Finnish partner as Nokia bought up all Symbian shares and then bequeathed its code to the Foundation. But the time was the essence here as the pressure from Android was relentless.

When Nokia announced its decision to acquire the rest of Symbian and offer Series 60 and Symbian OS software for free to the open-source community, the mode of Nokia's transition initially created some confusion. The Symbian Foundation—a not-for-profit organization setup to manage all the assets related to Symbian—would take a phased-in approach. First, it would make available the source code of Series 60 and Symbian OS for free to all members of the Foundation in the first half of 2009. During that time, Foundation members—who were asked to pay US$1,500 membership fee to join—could use Foundation's intellectual property assets but would not be allowed to redistribute them. By the first half of 2010, the Symbian Foundation was scheduled to be fully ready to go open source. It never came to that.

The truth of the matter was that Symbian needed to untangle copyrights and intellectual property rights to a massive collection of third-party commercial software—originally gathered and implemented by Symbian when it was a for-profit commercial entity. Uniting Symbian OS, the Series 60 platform, Motorola's UIQ and DoCoMo's MOAP (S) was also going to take some time—two years to completion was the target mentioned by the Foundation. Symbian also had to do a lot of scrubbing of its existing codes before the Symbian Foundation could completely hand off its software to open-source developers. Integrating Symbian's various software stacks into one unified platform and managing the millions of lines of codes were overwhelming tasks. So the transition could be costly, making Nokia vulnerable in that transition that could take around half a decade.

The old Symbian and the Series 60 platform had helped Nokia modestly succeed in the pre-iPhone days. Over the years, various incarnations of the Symbian OS had powered smartphones with desktop-like features such as preemptive multitasking, memory protection, and Unicode support. The biggest leverage that Symbian OS and Series 60 had was the large installed base. But now that Symbian partnership was dissolved, Symbian was run by the Foundation, and the system would be fully open-source, it became evident that Symbian must do more to succeed than just going open-source. According to software experts, there were four major areas where the Symbian Foundation needed to fix things: user interface, app-development environment, developer support and the leadership vacuum for the platform. In a nut-shell, Nokia had been fighting mostly on a device-by-device basis when the battle had shifted to the ecosystem level.

Unlike Android and iOS, which could start from a blank sheet of paper, Symbian had to carry the past that created a bit of slowness in the UI experience. Symbian proponents were confident that it wouldn't take much to fix the UI because the UI was just the presentation layer of the operating system. Next, Symbian accumulated a variety of app platforms that had become prevalent over the years. Symbian technologists hoped that Qt could help change that, making it easy to create apps that would work across all past and future Nokia devices. The Qt platform and library, which had been extremely successful in the desktop world, allowing developers to create cool apps, could help the new Symbian overhaul its app strategy by creating hardware and software coherence. Apparently, Nokia needed a much easier cross-platform API to attract more developers, so it bought Trolltech's Qt development framework. The problem was that the purchase came a bit late.

That much about user interface and app development environment! The developer community at large still considered it imperative that Nokia led the development of Symbian if the operating system was to succeed, much like what Google had done with Android. There were people in the industry who believed that the Symbian open-source experiment would fail and that Nokia needed to scrap the Symbian Foundation and bring the OS software home. "It would eliminate time-wasting Foundation activities like release councils, architecture councils, user interface councils," Gartner's London–based analyst Nick Jones wrote on his blog. "What Symbian needs is agility and vision, not committees, and if Symbian is fixable it will be fixed a lot faster under a single leader. Great user interfaces aren't developed by committees."

Jones asserted that if Nokia took Symbian back in-house, assumed control and leadership, and used Qt to create a good user interface, Symbian would be a very different entity from what it was in 2010. Other industry watchers saw Nokia's decision to acquire Symbian and turn it into a non-profit entity to leverage the collective power of the open-source community as an Android me-too. Merely opening one's operating system was in itself no guarantee of success on the mobile market. Moreover, after a decade of trial and error, time wasn't on Symbian's side, and Nokia needed to move fast. The Series 60 platform was still significant as of 2010, but that leverage could diminish in the coming years if Nokia failed to put the Symbian house in order.

The Symbian Foundation had been in a state of flux since its inception with the ongoing streamlining of the workforce. Ultimately, Nokia was to face the difficult decision of whether to abandon Symbian entirely and rebuild its software strategy all over again. There had been speculations for months about whether Nokia would adopt another platform before the world's largest handset maker embraced Microsoft's Windows Phone 7. Symbian loyalists said it wouldn't come to any of that: "The strategy is clear and makes sense." But Nokia's history with Symbian had been checkered. Its Symbian Foundation liaison now made the Finish wireless house look like it was between a rock and a hard place.

THE FALL OF SYMBIAN

In the hindsight, Nokia's decision to open source Symbian in 2008 to try to compete with Android was too little too late. The end game also exposed a fatal flaw in the Symbian anatomy: it was a phone-centric platform. Nokia engineers seemed to have the phone-first mindset that ran through everything they did. They put the phone function first rather than Internet-connected services because it was the phone that Nokia was selling like hot cakes. Their view of the mobile computing landscape was so clouded by the phone-first mindset that it inevitably hindered their ability to adapt to the evolving computing paradigms. That eventually led to a failure to recognize and deliver on the true potential of smart devices.

The Symbian saga offered some valuable lessons. First, one can't be all things to all people all the time. Second, if merely acquisitions and strategic partnerships mattered, then the alliance of Ericsson and Microsoft to create a compelling wireless e-mail service in the late 1990s would have rendered BlackBerry irrelevant. Third, Nokia ended up spreading resources thin over too many different projects. Fourth, timing was crucial in the software business. As Gartner analyst Carolina Milanesi put it, Nokia was guilty of "trying to fix Symbian for too long." Regardless of how technically powerful Symbian was, there was no getting away from the problem that it was legacy technology, built for an earlier mobile era when phones were phones first, not mobile computers.

Now how would Nokia come out of this software conundrum? The common perception among mobile industry watchers was that the world's largest cell phone maker would eventually dedicate Symbian to low- and mid-tier feature phone-like smartphones while steering its premium phones toward MeeGo, the OS platform it was co-developing with Intel. That made sense, at least on paper, as it could provide Nokia with much-wanted room to correct its missteps and reinvigorate its software roadmap. But mobile software technology probably grew faster than Nokia's ability to handle it. The epic tale of the world's first smartphone platform came to a sad end in February 2011 when Nokia's new captain concluded that it would take too long to modernize Symbian, and pulled the trigger on the ten-year-old software, making it a franchise platform.

Symbian had proven to be non-competitive in leading markets like North America. It was also proving to be an increasingly difficult environment in which programmers could effectively develop apps to meet the continuously expanding consumer requirements, leading to slowness in product development and creating a drawback when Nokia sought to take advantage of new hardware platforms. Stephen Elop called spade a spade before making Windows Phone 7 the primary operating system for Nokia smartphones and integrating his company's software services into Microsoft's. It's ironic that like Nokia, Microsoft had also struggled in the smartphone arena. And just like Nokia, Microsoft had been a victim of its own success. Both companies stayed with their playbooks for too long and didn't change with the times.

Symbian managed to keep its head above the water and continued to be updated to service the millions of existing users through a franchise platform handled by Accenture. But the overall outcome was an immediate plunge in sales of Nokia smartphones that crushed the Finnish company's balance sheet. What pulled Nokia down was not as much the choice of strategy, but how the changeover from Symbian to Windows Phone had been implemented.

If Nokia had vowed to offer both Symbian and Windows Phone devices without putting a time limit on the older operating system, it could have retained the market and sales that slipped away faster than originally anticipated. By publicly stating that Symbian was a dead end, Elop might have shaken an already fragile public perception of the cast-aside Symbian software. On the flip side, however, if Symbian continued along the path where it underwhelmed, Nokia could have run the risk of having a negative vibe to its transition to the Microsoft-powered phones. It was also plausible that Microsoft might have wanted Elop to make this call as part of the overall deal simply to avoid a scenario of contagion.

What happened next was that the Symbian portfolio went into a terminal decline, and unbranded phones began sapping Nokia sales in critical markets like China and India. Nokia's expectation of Symbian-based smartphone sales slowly descending as Windows Phone took over was measured in years, but the market responded in months. Nokia was now in a tight corner. It was being pinched from below with the dying Symbian platform and from above with the unproven Windows Phone. But the Finnish device maker had to live with this miscalculation and remain committed to its larger smartphone strategy.

Symbian continued fading away all through 2012 while Nokia, along with Windows Phone-based Lumia, began to cross promote lower-cost Asha handsets made on its S40 proprietary operating system software. Symbian OS software platform was officially dead in January 2013 when Nokia announced that its 808 PureView handset—the super camera phone which came to market in mid-2012—was going to be the last Symbian device from Nokia.

MEEGO HOPE FLOATS

In February 2010, at the Mobile World Congress held in Barcelona, Nokia startled the technology world by announcing the launch of a Linux-based open-source operating system in collaboration with Intel Corp. In the midst of a highly draining exercise of first taking control of Symbian and then turning it into an open-source undertaking, Nokia had decided to also adopt a separate operating system named as MeeGo. The world's biggest chipmaker and the world's largest mobile handset manufacturer were joining forces around a Linux operating system to create an über-platform for the next generation of computing devices: smartphones, pocket computers, tablets, netbooks, and more.

Nokia executives told a stunned crowd that through open innovation, MeeGo would create an ecosystem that would be second to none, drawing in players from multiple industries. They asserted that due to the spread of cloud computing and new advances in electronics and networking technologies, mobile devices would increasingly move beyond smartphones to include other computer-like gadgets such as tablets, and here the MeeGo platform would be an important asset for Nokia. They also affirmed Nokia's commitment to the Symbian platform, saying that MeeGo was conceived as a base for a wide array of new computing-centric devices given that market for handheld gadgets was deftly diversifying. MeeGo was a Linux-based software platform designed to work across a range of hardware architectures and devices including mobile computers, netbooks, tablets, media phones, connected TVs, and in-vehicle infotainment systems.

However, the story behind the story was that Nokia had recognized back in 2007 that its Symbian operating system was increasingly growing obsolete. The Finnish mobile phone giant had heavily relied on Symbian, which enjoyed popularity in terms of market share, but was saddled with an archaic and needlessly complicated interface that hadn't adapted well to the world of touchscreen-based next-generation smartphones. In 2008, Nokia senior executives had come to acknowledge that matching Apple's slick operating system amounted to their biggest challenge. So while one team tried to revamp the aging Symbian, another effort dubbed as Maemo attempted to build a brand new operating system from the ground-up. Nokia had been working on this Linux-centered software platform for conceiving new-age smartphones and tablets since 2005.

In 2009, Intel had also started a project called Moblin for creating a Linux-based operating system initially designed for netbooks. However, Intel, while being successful in supplying its Atom chips to the netbook market, hadn't made significant inroads into the smartphone domain. The silicon bellwether was hoping that a modern operating system platform might help it leverage its chip business in the rapidly growing smartphone market. There had been a number of theories as to why Intel and Nokia decided to unite their Moblin and Maemo mobile software platforms.

Two trends especially shaped the need for MeeGo. First, the ubiquitous Internet offering constant connection regardless of location and, second, the ability to access that connection through a variety of devices: in one's pocket, in the car or kitchen, in the living room, through the television, and so on. No single device would fit every need, but they would all be connected in some way. But more than anything else, this marriage underpinned the fact that both companies were serious about being leaders—and survivors—in a mostly open and truly flexible mobile computing ecosystem. Intel and Nokia wanted to work together to make the best mobile computers in the world. Both companies desperately wanted to claim a stake in the next-generation mobile computing market. Nokia, specifically, was looking to boost the competitiveness of its smartphones as well as widen its presence in other device areas through the MeeGo software.

During the late 2000s, the rise of smartphones and the growing popularity of tablets and streaming media players had opened the doors for the new operating systems that promised a better user experience. Take Android—launched in 2008 for smartphones—which later spread to tablets and even birthed Google TV, a platform that combined cable TV programming with websites from the Internet. But why did Nokia pick the computing powerhouse Intel as its new partner? The reasoning probably laid in the fact that, while geared for mobile devices, Android and iOS, two of the most successful operating systems, stemmed more from a personal computer lineage compared with Symbian's mobile telecom roots. The dramatic swing on Nokia's part befitted the new era of mobile devices that resembled small general-purpose computers more than single-purpose phones.

Customers now used the phone not just for calls and text messaging but also for e-mail, web browsing, and games. They added new software picked from application stores rather than using a limited list of applications supplied with the phone. Nokia was essentially positioning MeeGo as its go-to platform for top-tier devices while Symbian would be used for its feature phone-like low- and mid-tier smartphones. Symbian wasn't supposed to compete with iOS and Android; that would be MeeGo's job. But the MeeGo story came to a crossroads just when the nascent platform was starting to become the collective hope of the Finnish company. In February 2011, Nokia's newly arrived CEO, Stephen Elop, sent shockwaves across the industry by going back to his former employer, Microsoft, in a bid to form a combined front against the iPhone and Android.

Although Nokia initially pledged its commitment with MeeGo and continued spending on its development along with Intel, developers were most likely to abandon MeeGo just like Symbian. It was already an uphill battle for MeeGo because it wasn't about developing OS software anymore; the fight for the soul of the smartphone had turned into a battle of ecosystems. The collaboration between Intel and Nokia had failed to produce a phone in its first year. Jo Harlow, in-charge of Symbian software operations, told her boss Elop that MeeGo wasn't mature enough and that Symbian could not fill the gap. By that time, Nokia was on track to introduce only three MeeGo-based models before 2014, and that was far too slow to keep the company in the smartphone game. Elop also knew quite well that a half-backed software platform like MeeGo could cost even more.

It's pretty ironic that the Nokia N9—the first MeeGo-based device—was one of the most fascinating phones of the last few years. The MeeGo-powered N9 had a lot of love in tech circles. Still, it was effectively dead on arrival when launched in September 2011. In retrospect, the MeeGo saga symbolized Nokia's muddled smartphone strategy and its overly risk-averse management style. The fact that Nokia began developing MeeGo software at a time when it still called Symbian its principal OS platform just showed that the Espoo, Finland, company was far from clear in its long-term commitment to either platform. Moreover, by attempting to juggle both Symbian and MeeGo, Nokia showed that it didn't understand the importance of ecosystems.

Nokia's selection of Windows Phone had practically marked the end of the road for MeeGo software. Intel, apparently unhappy with this about-face, initially vowed to find new partners and carry on. However, given the transformation of mobile OS software into a vibrant ecosystem and the consolidation that would come as a consequence, such efforts might not stand much of a chance, even if an industry giant was behind them. Having let down by Nokia, in September 2011, Intel decided to hedge its bets with Google's Android platform owing to the sheer number of people in the ecosystem. Intel cobbled a team of developers to make Android apps run more seamlessly on its Atom hardware platform and thus renew challenge for the previously untouchable ARM chip ecosystem.

MEEGO REBOOTS AS JOLLA

Moreover, in September 2011, after MeeGo was scrapped, Intel joined the Linux platform for mobile (LiMo) in collaboration with Samsung and the venture was renamed as Tizen. The LiMo Foundation was formed in 2007—a year before the launch of Android—and it quickly lined up big-name members such as Samsung and Vodafone to create a Linux-based open platform. It gained further momentum when heavyweights like Verizon Wireless and SK Telekom joined the mobile initiative and committed in a big way. However, while Android flourished, LiMo floundered as a half-baked project and was eventually abandoned in 2011. Tizen drew its heritage from the aborted MeeGo initiative, the resurrected LiMo platform and Samsung Electronics, who had rolled its homegrown Bada mobile OS software into Tizen.

Samsung and Intel would steer the development of Tizen operating system that they claimed to be more open and customizable than Android. A steering group made up of employees of Intel and Samsung would carry out the development work while other members of the Tizen Association would be allowed to contribute code and suggestions. Tizen proponents claimed that the new platform would offer a better battery life and support for wearables and TVs. Well aware of the app development dilemma, the open and customizable OS platform focused on fully supporting HTML5, though Tizen also had the option of native apps for more-sophisticated programs.

On the other hand, for many Nokians, MeeGo seemed something to fall back on. They still saw signs of life in this Linux-based software platform. So, almost parallel to the making of the Tizen, in October 2011, Nokia spun off the MeeGo software platform as Jolla through its Bridge Program to let the new venture build a fully-featured operating system. Bridge program provided former Nokia employees help in the form of technical training, financial aid, and in some cases, patents. Jussi Hurmola and Marc Dillon assembled a team of around 50 employees, many of whom had worked on the Nokia N9 handset. Jolla Oy, a Finnish startup run by former Nokians, mostly from the MeeGo team, had no financial ties with Nokia. But spiritually it was still a Nokia spin-off.

Nokia was going through a massive transition in 2011 with lots of layoffs; so the upstart Jolla was able to recruit key talent, including engineers that had worked on the MeeGo project. This crack team of ex-Nokians had a good idea how large the market was and knew what was attractive in terms of hardware, software, and technical support. Take one of the company CEOs and co-founder Marc Dillon. He had worked at Nokia for almost eleven years and had played a significant role in bringing out the well-reviewed N9 handset.

Jolla meant a one-man sailor boat in Finnish. The name symbolized the idea of big sea with a great opportunity; sailing alone could be adventurous and a bit dangerous, but this group of former Nokia employees wanted this adventure because they saw a great opportunity. The crew of fewer than hundred Finns, most of them refugees from the turbulent Nokia ship, kicked off the venture with a true startup spirit and tasked themselves to develop and nurture a legitimate alternative to the Coke and Pepsi of the smartphone world: Apple's iOS and Google's Android.

Jolla made waves by announcing it was focusing on China, a market that was already churning out cheap Android mobile phones. China was the most dynamic smartphone battleground where, by 2012, Android had dominated by capturing nearly 80 percent of the market. Apple's iPhone stood out being different, but it was also hugely expensive for the Chinese smartphone buyers. The Finnish mobile upstart Jolla planned to kick-start from China also because it was the largest market producing first-time buyers of smartphones fast enough to give this revolutionary idea a fighting chance. In order to get its operating system, and eventually, Jolla-branded phones in front of enough Chinese smartphone buyers, the Finnish firm had announced a partnership with the largest mobile chain retailer in mainland China: D.Phone.

The first Jolla-based device would be launched in China in summer 2013 through the distributor network of D.Phone which boasted 2,000 stores. In China, things worked a little differently than they did in western mobile phone markets. Here, the retailers bought handsets and charged customers full price for them. The way wireless operators competed for exclusivity was to offer retailers and their customers the biggest possible amount of free airtime. So retailers like D.Phone wielded enormous influence over the mobile phone market.

But Jolla wasn't content on merely recycling MeeGo operating system software for large smartphone markets like China. Nokia's spinoff Jolla was grappling to create a brand new mobile platform with a mass market appeal. To broaden the horizon and industry participation around the new ecosystem, Jolla decided to build an alliance named "Sailfish" around the OS platform and make it ready for licensing by spring 2013. The alliance would be based in Hong Kong and would have a wider scope than the Jolla phones. Sailfish, for instance, could serve a greater variety of electronic devices including automotive, smart TV, and tablet computers. In a nutshell, Jolla intended to build up as a smartphone maker with a total of about two hundred employees. The rest would lie in the ecosystem.

Jolla based the Sailfish software platform on the swipe gestures, a convention borrowed from the hugely popular Nokia N9 handset. The N9 handset was one-of-a-kind phone which Nokia released in September 2011 to kick-off MeeGo, the direct ancestor of Sailfish. The platform was designed to acknowledge that most people used their phones for several things at once. The phone's home screen could be filled with up to nine concurrently running applications. Swiping from one side of the tile or the other would, for instance, skip tracks if it was a music player, or flip through contacts if it was the address book. Ironically, here Sailfish resembled the Windows Phone, which had a home screen full of live tiles displaying information from various apps. But the tiles in the Windows Phone software only served to open the app when users tapped them.

Jolla was quietly taking its intriguing alternative to Android and iOS to one of the fastest growing corners of smartphone techdom: China. In a way, Nokia's MeeGo technology was aiming to disrupt American smartphone platforms by establishing itself with Chinese phone manufacturers. It's worthwhile to note that, in 2012, the situation had started to resemble what came to pass in the PC era—the underlying control of the mobile market was now in the hands of American software players. So an indigenous mobile platform like Sailfish, one that was not dependent on an American technology, could potentially attract the Chinese mobile sector.

THE CURIOUS CASE OF SAILFISH

The two key challenges for Jolla's revitalized mobile operating system Sailfish were ease of use and market reach. A software platform required a critical mass of either users or developers to make an impact in the market. The brief history of smartphone had proven that it was a winner-take-all market. Smartphone users generally cared less about who made their phone and more about whether or not it would run the apps they wanted. As a result, developers wanted to create apps for the platforms with the most users. Moreover, Apple and its fast follower Google, which pioneered the smartphone app store, had made it too hard for newcomers to enter the market.

Just ask Microsoft. The Redmond, Washington–based IT behemoth had pumped massive resources into its Windows Phone ecosystem to make it apps-friendly and had only achieved fair results. According to media reports, Microsoft had spent five to ten billion dollars and still hadn't been able to compete two years after it introduced Windows Phone software. A smartphone without apps was going to be undesirable, so building an app ecosystem was the biggest challenge for Jolla. Jolla's young Finns at the helm were well aware of how hard it was to break into the field of entrenched giants. They thought they had built a strategy to counter this challenge. Going to China was just the first part. Next, they wanted to develop an interpretation layer on top of Sailfish, which allowed Jolla phones to run most Android apps. Third, the nascent firm was trying hard to create a culture that would attract open-source movement.

The Jolla leadership claimed that Sailfish would always be open-source and that the company would make money by licensing the patents Jolla had on its user interface. The Finnish startup firm—mostly runs by software developers—espoused open-source values like sharing, collaboration and transparency. Jolla managers hoped that developers would contribute to Sailfish software code for free in their spare time, as typically happened in open-source projects.

Jolla, on its part, was aiming for an ecosystem open and friendly enough so that a handful of dedicated HTML5 and Qt developers and enthusiasts could fill in the gaps and get more out of the device. Android had commanded some of that open-source goodwill in its early days, but then Google had to use heavy-handed tactics to keep Android handset manufacturers in line. Jolla was pinning hopes on exploiting those holes in Android's open-source credentials. In fact, through that specific premise, the Finish startup was aiming to launch an attack on the heart of Android business model.

Take Samsung, the leading producer of Android handsets, as an example. The Korean mobile device maker had to obey the rules laid down by Google so that it could feature the Android app store—Google Play—on its devices. Samsung couldn't use another app store. Moreover, revenue from

Google Play went to Google, not Samsung. By contrast, if a mobile operator or manufacturer wanted to launch its own app store or music service on the Sailfish platform, it would be fine. Jolla aimed to have less control over Sailfish than any other OS maker had over its software platform. The Sailfish alliance encouraged handset makers, software companies, and mobile operators to have a stake in the success of this embryonic platform by digging into the software code. They could make suggestions, tweak software for local markets, and put in new services that could help them generate revenue.

Another potentially crucial advantage of Sailfish, in comparison to Android, was freedom from the dreaded patent wars. Both Nokia and Intel had made massive investments to keep Sailfish's predecessor MeeGo clear from intellectual property issues. MeeGo operating system had been written specifically to avoid infringing the existing patents. On the other hand, a closer look at Android's peculiar business model and patent disputes that usually troubled Android device makers revealed that there was a de facto licensing fee for mobile operators and manufacturers despite the fact that Android was free software. In a strange about-face, Sailfish—the descendent of MeeGo software platform—was now aiming to disrupt Android that had brought down Nokia's 800-pound gorilla Symbian just a while ago.

Sailfish, a revamped version of the software platform MeeGo, claimed of being faster and easier to use than Android, iOS and Windows Phone. However, while the round two of the pioneering MeeGo work had started, it was apparent that the battle was going to be hard and long. The Finnish upstart Jolla had a very steep road to climb to get to market. The first roadblock came in summer 2013 when Jolla missed its much-talked about phone launch in China. The promising start eventually turned into a vague plan when Jolla announced that both widespread distribution plans in China and US$260 million in backing for the Sailfish Alliance were on hold.

On November 27, 2013, the boutique smartphone manufacturer sold 450 handsets in a single day sale at the Finnish mobile operator DNA's a pop-up shop in Helsinki. The plucky smartphone startup housed at a stone's throw away from Nokia's headquarters in Espoo had finally kicked off its journey by partnering with the Finnish mobile operator DNA. In the early going, Jolla handsets could be purchased SIM-free in Europe as opposed to being widely available through a network contract. And the initial goal was to sell 100,000 units.

Later, in 2014, the Finnish upstart expanded Jolla phone pre-sales to three European countries: Britain, Finland, and Germany. Next up, Jolla expanded distribution of its handsets through mobile operators in Estonia, Hong Kong, and Kazakhstan. Shortly afterward, the independent phone maker from Finland inked a deal with India's largest online marketplace Snapdeal to take Jolla handsets to one of the world's largest potential markets for smartphones.

The Jolla hardware looked similar to that of Nokia's Lumia with a clean, button-less front face that housed the 4.5-inch touchscreen. And Jolla's phone priced at around US$535 was considerably more expensive than handsets with similar specifications. Apparently, to make inroads in a highly competitive smartphone market, Jolla was positioning itself as a boutique smartphone shop. In other words, as Ewan Spence wrote in his post on Forbes.com, Jolla wanted to become the Ferrari of the smartphone world. That was the antithesis of what was going on during the mid-2010s in the smartphone industry that had been largely focused on the volume side. The key smartphone players were busy in grabbing market share from others and building the biggest app stores.

The eighty-plus startup of ex-Nokians who made the decision to carry on developing a platform that Nokia abandoned had a come a long way. Still it was a long road for Jolla because it wasn't merely a smartphone maker; Jolla wanted to turn one phone into an open mobile movement. It was just the beginning of a challenging journey for the boutique smartphone maker located at a place in central Helsinki that had once housed Nokia's research lab. Jolla acknowledged the harsh market realities and the fact that being a small company it had to proceed in phases.

Jolla co-founder Marc Dillon told the press that the company had been talking to almost everybody in the mobile space over the past two years. "If you just look, people wise, then we're really small compared to these guys. Their coffee budget is probably as much as we've spent to create this device!… So now they have a way to judge our capabilities by taking a real device and looking at it, and seeing what it does."

Nokia executive vice president of new devices Kai Oistamo (right) and Intel senior vice president and general manager of software and services Renee James announced the launch of MeeGo platform to a stunned crowd at the Mobile World Congress in Barcelona on February 15, 2010.

Photo courtesy of *Reuters*

Could Jolla be a MeeGo savior? A group of ex-Nokians and MeeGo enthusiasts who formed this small enterprise thought so. Three of the five Jolla co-founders—Stefano Mosconi (left), Marc Dillon (middle) and Jussi Hurmola—are seen here wearing the badge of Nokia's wonderful, yet doomed, MeeGo software platform.

If there was one product that symbolized Nokia's comedy of errors, it was probably N9 handset. The MeeGo-based device was one of the most fascinating handsets in years; according to some industry watchers, it was the world's most underrated smartphone. The high demand for N9 phone brought a critical dilemma for Stephen Elop, who had already announced to shut down the MeeGo team after releasing N9. He desperately wanted Nokia's partnership with Microsoft to succeed. Elop put a closure to the future of N9 by asserting "There is no returning to MeeGo, even if the N9 turns out to be a hit."

Image: Nokia

4 NOT SMART ENOUGH

"Oh my God, we had it completely nailed. I was heartbroken when Apple got the jump on this concept, when people say the iPhone as a concept, a piece of hardware, is unique, that upsets me."

— Frank Nuovo, the former chief designer at Nokia, who had demonstrated a mobile Internet-capable touchscreen handset to wireless operators and investors back in 2000

Nokia's Anssi Vanjoki first heard about the Internet in 1993 at a time when his company was heavily betting on digital cellular phones. One day Vanjoki saw a new hire hunched over a strange-looking database on his PC. It turned out the newcomer was online, using the Gopher menu system to browse through a library at the University of Texas. Vanjoki thought that if he could do this on the

PC, he should be able to do the same on a digital handset. Across Scandinavia, many engineers were having similar thoughts. It might have seemed implausible at that time that people would surf the Internet and exchange e-mail messages from their cell phones given their tiny screens and awkward keypads. However, about six years later, mobile phone users were unmistakably doing so and in reasonably large numbers.

No one had embraced the idea of wireless web more enthusiastically than Nokia. "In ten years time, I would like Nokia to be dubbed as the company that brought mobility and Internet together," said its CEO Jorma Ollila back in 2000. "It's not going to be easy, but this organization loves discontinuity; we can jump on it and adapt. Finns live in a cold climate: we have to be adaptable to survive." After the advent of digital wireless telephony, Nokia caught the GSM tsunami in the 1990s and gained a formidable market position. The next wave was the mobile Internet and Nokia's future seemed to depend on making the right call for the next generation of data phones.

But the drive to the mobile Internet was pushing Nokia from the simple, radio-based handsets it knew so well into the dizzying world of computers and consumer goodies all of which would communicate over the wireless Internet. Here, at this very crossroads of digital convergence, Nokia was to smack into powerhouses like Microsoft and Sony, who were making inroads toward similar goals. Data was new, and that changed everything. Mobile phone firms were first taken off the guard by the surge in voice services; they were surprised even more by the success of prepaid phones and text messaging. While struggling to meet demand, wireless makers and operators could do no wrong as very little marketing was needed for a voice-centric environment. Next-generation mobile services, particularly those using data, were far less likely to be successful by a similar fluke.

A report published by Nomura, an investment bank, questioned Nokia's ability to manage the transition to wireless data. "Is Nokia the IBM of our times," it asked back in 2001. The challenge that Nomura had pointed out was real, though the analogy looked flawed at that particular time. IBM became ossified after sitting atop its industry for decades. Nokia, by contrast, was no stranger to reinventing itself. Nevertheless, Nokia heeded the call by evading the "Nokia-against-the-rest" temptation that could translate into unavoidable technology risks. The market predicament led to the formation of the Symbian alliance, which in many ways was reminiscent to the inception of Wireless Application Protocol (WAP)-based mobile Internet initiative, owing its life to the industry clout of the wireless triumvirate made up of Ericsson, Motorola, and Nokia. And Motorola was a Johnny-come-late during the formation of Symbian just like the WAP affair.

Naturally, the top three wireless makers wanted to stay in control and keep all the money in the family. However, with the heyday of the 1990s behind them, the wireless troika was facing a strong competition from handset makers in Japan and South Korea. The smartphone was indeed a new class of gizmos, and though it resembled ordinary cell phones, that didn't give Nokia and other leading handset makers a birthright to the market. Although, in the early going, the mobile Internet run was seen as about to sweep the wireless industry by incumbents like Nokia rather than outside Internet competitive forces. But to lay claim on the mobile Internet, these handset incumbents were stretching far from their radio roots, and that quantum leap in technology was adding to the uncertainty.

Smartphone and mobile phone continued to converge and collide during the mid-2000s, and when the second act in the smartphone game got underway later in the decade, Nokia turned into a sad story because of its loss of leadership. At one stage during 2007, the Finnish mobile giant had garnered 40 percent of the global handset volume, according to research from Strategy Analytics. Shortly afterward, however, the company that embodied wireless industry's quest to inherit the efficiency of PC was seen as turning into an also-ran. How on earth did that happen?

Although Nokia's focus remained on mobility and the company had no intention to extend its resources to the traditional computer business, Nokia understood quite well that fundamental changes were happening in the computer industry. To its credit, Nokia foresaw a day when wireless handsets would be as powerful as computers. But mobile handset makers like Nokia, coming from a telecom background, prioritized hardware muscle and superior specifications, while computing industry players such as Apple and Google emphasized user interface and software to enhance ease of use. Nokia managers avidly talked about multimedia mobile computers, but they couldn't look beyond Communicator-like phones with twelve keys. Now, being surrounded by competitors of the PC origin, Nokia seemed to have lost the battle for smartphone dominance, even the race to be a runner-up to the iPhone.

NOKIA PROCRASTINATION THY NAME

In 1999, a *New York Times* profile story had credited Nokia product designer Frank Nuovo for "turning cell phones into fashion statements." Now Nuovo was saying that the Finnish device maker had prototypes that anticipated the iPhone more than seven years before Apple actually launched its game-changing mobile handset. A Nokia design team had demonstrated a phone with a color touchscreen set above a single button. The phone could locate a restaurant, play a racing game, and order retail items like cosmetics and food. There were also media reports suggesting that, during the late 1990s, Nokia had developed a tablet computer with a touchscreen and a wireless connection, the features that became the hall-mark of Apple's iPad tablet computer.

"It was very early days, and no one really knew anything about the touchscreen's potential," recalled Ari Hakkarainen, a manager responsible for marketing in the development team for the Nokia Series 60. "And it was an expensive device to produce, so there was more risk involved for Nokia. So management did the usual. They killed it." Nokia introduced the industry's first touchscreen devices in 2003—the 6108 and 3108 phones—that worked with a stylus. Later, in 2004, Nokia launched its first full-fledged touchscreen. The 7710 handset had usability and stability issues and was not a success. In retrospect, the Finnish phone giant did not perfect the technology to fingertip precision before Apple did.

The company's in-house reviews showed that people were not excited about touchscreens. Since product programs at the Finnish phone giant tended to avoid risks in general, Nokia managers decided to drop touchscreens devices. The decision, according to *Helsingin Sanomat*, was made in 2005 and savings were used to make non-touch products even better. Not surprisingly, therefore, Nokia refused to jump on the touchscreen bandwagon and waited a year after the iPhone launch to unveil its first touchscreen phone: the Nokia 5800. More so, Nokia 5800 was less a smartphone and more a handset optimized to play music.

There was more anecdotal evidence showing how Nokia developed the sort of devices that consumers actually wanted, but the Finnish company wasn't able to bring them to market. Kai Nyman, a former architect for Internet services at Nokia, had known about the Internet-ready touchscreen phone way back in early 2000s. He pitched this idea to the management, which was reluctant to proceed because of concerns over the performance of Symbian operating system. In 2004, the prototype was demonstrated to business customers at Nokia's headquarters in Espoo, Finland, as an example of what was in the company's pipeline. The management didn't pursue development, worrying that the product could be a costly flop.

Ari Hakkarainen also recalled that his team developed the early design for a Nokia online applications store in 2004. They made a demo and tried to convince middle and upper management, but there was no way. The company rejected an early design for a Nokia app store—an innovation that Apple introduced four years later, and other handset makers quickly followed the suit.

Similarly, Juhani Risku, who worked on user interface (UI) designs for Symbian from 2001 to 2009, remembered hundreds of proposals that his team offered to improve Symbian but could not get even one through. In this gigantic software chaos, according to some industry estimates, Nokia had wasted 2,000 man years on UIs that didn't work. "It was management by committee," said Risku, comparing the company's design approval processes to a "Soviet-style" bureaucracy. Ideas fell victim to infighting among managers with competing agendas, and were rejected as too costly, risky, or insignificant for a market leader that had become synonymous with mobile phones. Risku later left Nokia to design environmentally sound buildings.

His account of Nokia's highly staid corporate culture was a testament of the struggles shared by Nuovo, Nyman and Hakkarainen for getting Nokia's smartphone campaign back on track. "There were plenty of years to make Symbian better," Nyman asserted. "We could have rewritten the whole code several times over. We had the resources and the people. But we didn't do it." Now even Apple's co-founder Steve Wozniak didn't hesitate to take a stab at Nokia while commenting about the rivalry between Android and iPhone. He called Nokia a throwback from a previous generation. Wozniak told Dutch newspaper *De Telegraaf* that "Nokia came late to the touchscreen game and now has 'an image problem' that it needs to rectify with a fresh new brand and younger customer base."

Nokia had grown too fat with an exponential growth in the mobile phone market. It took an awfully long time for the Finnish handset maker to set a strategic direction and execute it in an efficient and timely manner. The organization structure became so convoluted that it became hard for Nokia to create phones with coherent, consistent, and beautiful design. The problem of a number of groups competing with each other for support within the company and the attention of top executives plagued most of the innovative initiatives at Nokia.

However, in fairness to Nokia, when dealing with a gigantic manufacturing machine that pumped out millions of phones, a single mistake or bad call could cost the company billions of dollars. As a result, management was structured around many layers of meetings and approval bodies. The whole structure was built to prevent mistakes. Nokia was evidently good at coming up with innovative new concepts to show the world what the future's going to look like. Meanwhile, it was selling feature phones with moderate Internet capabilities in huge volumes, making massive profits, keeping its shareholders happy, and clamoring for more of the same.

The mobile phone maker wasn't thinking big enough when it really counted and was content with its blue sky pursuit. In fact, it was caught between the two worlds: feature phones and smartphones. That led to an emphasis on incremental innovation of existing products rather than aggressive push toward disruptive innovations like the smartphone.

It underlined the fundamental dilemma for a company that had long excelled at making beautiful mobile phones. Ultimately, the success in feature phones made it harder for Nokia to change its business to react to the looming threats from Internet-focused companies.

Qualcomm chief Paul Jacobs recollected his experiences of working with Nokia in an article titled "Nokia's Bad Call on Smartphones," published in *The Wall Street Journal* on July 18, 2012. "What struck me when we started working with Nokia back in 2008 was how Nokia spent much more time than other device makers just strategizing," he recalled. "We would present Nokia with a new technology that to us would seem as a big opportunity. Instead of just diving into this opportunity, Nokia would spend a long time, maybe six to nine months, just assessing the opportunity. And by that time the opportunity often just went away."

A sense of denial was another problem. Nokia engineers would look at the smartphone market and say "we invented all that stuff." They just couldn't see that mobile users were not using Symbian phones to go online. The internal reports from Nokia engineers emphasized that the iPhone was expensive to manufacture and only worked on second-generation GSM networks; Apple offered support for 3G networks a year after the iPhone roll-out. An internal report also noted that the iPhone didn't come close to passing Nokia's rigorous "drop test," a hardware benchmark in which the handset was dropped five feet onto concrete floor from a variety of angles.

Nokia's liaison with software developers—who have a keen insight of how fast or slow a platform is gaining traction—provided another vantage point. By 2010, they were flocking to competing platforms—Android and iOS—despite the fact that Nokia was still the top smartphone maker in terms of market share. After finding out that software developers were not heeding Nokia's calls for building apps on Symbian platform, the Finnish device maker decided to streamline the process to make it easier for developers to write apps for Nokia smartphones. For a Silicon Valley company, such decisions could have been made around a conference table. But for Nokia, about hundred engineers and product managers from around the world—from Massachusetts to China—had to gather in a hotel ballroom in Mainz, Germany to hash out the details. They took three days while representatives from Symbian, MeeGo, and other software projects struggled to make themselves heard.

It'd also be worthwhile to mention that Anssi Vanjoki—the guy in charge of Nokia's Mobile Solutions unit—was spearheading the MeeGo operating system project with 2,000 people serving under him. That was in a stark contrast to Google that managed to get better results with some 500 Android engineers. Vanjoki, one of the two internal candidates for the top job at Nokia, was the company's smartphone guru. According to the Finnish news site Kauppalehti, Ollila actually wanted to put Vanjoki in charge instead of Stephen Elop. Vanjoki left Nokia soon after the announcement of Elop replacing Olli-Pekka Kallasvuo in 2010 and became professor at the Lappeenranta University of Technology in Finland.

Later, in October 2012, Vanjoki became the chairman of ultra-high-quality handset company Vertu. The luxury mobile phone maker, formerly owned by Nokia, was eventually acquired by EQT. Vanjoki had joined Nokia in 1991 as vice president of sales in the mobile phones unit. During the 1990s, Vanjoki had devised Nokia's successful branding strategy and had helped choose the segment of the Spanish guitar waltz Gran Vals to become the Nokia ringtone. Moreover, he had the success of the N95 handset under his belt. It was also Vanjoki who had predicted that camera phones would replace digital single-lens reflex (DSLR) cameras. At the same time, however, he carried a lot of baggage in the failure of Symbian and MeeGo software ventures.

NOKIA VERSUS NOKIA

During the early 2000s, when smartphones were in their infancy and when most mobile phones started to look the same, Motorola was able to break the status quo with Razr, a slim, slab-like clamshell phone with a larger screen, a stylish and flat keyboard, and a built-in camera. The Razr was a design marvel and an ultimate feature phone that tempted users with glitz and gloss—though often at the expense of functionality. After launch in 2004, the Razr sold more than 100 million units and earned the accolade of the "iPod of mobile phones." That led to criticism on Nokia that it was spending too much effort on high-end smartphones while its rivals ate into its lucrative business by selling relatively inexpensive fashion phones to upwardly mobile people around the world.

At that time, when consumers across the globe were happily buying Nokia's steady menu of candy bar-shaped cellphones, people in North America began eyeing flip phones, handsets with a clamshell design. Motorola, mounting a comeback of sorts, led the charge for flip phones and cemented its reign in North America as a top seller for nearly three years. Still, Nokia wanted to focus on expensive, high-end mobile phones, but wireless carriers wanted inexpensive models instead. They also complained that Nokia did not sufficiently tailor handsets to their needs. Mobile operators—important customers for Nokia—had started to cut back on purchases. Especially, the U.S. wireless carriers were increasingly looking for mobile phone makers to supply customized handset, a request that upstarts like Samsung and LG were all too happy to fulfill.

Korean handset makers could deliver phones faster, and they were able to pick up on Nokia's limitations. Take, for instance, the N95 handset, one of the Finnish company's ultimate showcase phones, which was largely ignored in the United States because operators refused to sell it. It was at this point that Nokia largely abandoned the U.S. market. Once the U. S. mobile operators backed away from Nokia, the Finnish company maintained a niche presence in the United States through a handful of boutique stores. In retrospect, Nokia's decision to abandon the U.S. market didn't have any immediate consequence. Nokia seemed less concerned about its lack of presence in the North American markets because it was betting on feature phones to capture the rising demand from non-U.S. markets. Nokians were quick to point to their market leadership as proof that they were still in a strong position.

In 2006, when Olli-Pekka Kallasvuo, Nokia's chief financial officer, took charge from Jorma Ollila for the top job, he merged Nokia's smartphone and basic phone operations. Subsequently, the corporate focus gravitated toward traditional mobile phones. It was a strategic blunder that shifted Nokia's focus from smartphones back to basic phones just when the iPhone was about to append the market. The Finnish mobile phone maker continued to gain market share around the world and hit its peak in 2007—the year of the iPhone launch. It's pretty ironic that "the iPhone moment" brought an irrevocable and dramatic leap forward to the sleepy smartphone world in 2007, the year that was also the best ever for Nokia's business.

That year, the Finnish company sold 436 million mobile phones, which was around 40 percent of the global handset market. Nokia's nearest contestant was Motorola who sold 164 million handsets. Nokia made a record profit of 6.7 billion Euros in 2007. No wonder, like BlackBerry, Nokia didn't take the iPhone's 2007 debut seriously. Nokia managers claimed that the iPhone battery was rubbish and that no one would use it once they figured that out. So, while consumers bought up all the bad-battery phones that Apple could make, Nokia managers had their heads firmly stuck in the sand. The irony was that Apple didn't actually know how to make mobile phones and had to learn it through a gradual and deliberate evolution. Case in point: Apple maintained carrier exclusives for two years. That gave Nokia and BlackBerry every opportunity to regroup and respond.

However, at around the same time, Nokia's grip had begun sliding even on feature phone business despite the fact that it was spending a lot of engineering resources on perfecting a variety of feature phones for the global market. Take the advent of the dual SIM mobile phones as an example of how Nokia missed the cues for key opportunities. Dual-SIM operation essentially enabled mobile consumers to use two operator services without carrying two phones. Using two SIM cards allowed a mobile user to take advantage of two different pricing plans for calls and text messages as well as mobile data usage. Nokia's close relationship with wireless carriers led to a sense of ambivalence, and as a result, the Finnish handset maker took a lot of years to embrace this trend. Nokia stayed away from the dual SIM bandwagon out of misplaced loyalty to large wireless operators who apparently preferred customers to use one network exclusively.

That crucial intersection point led to a lack of a sense of urgency because Nokia's share of both the smartphone and feature phone markets was in decline, but the drop-off wasn't dramatic. More important was the fact that the market share illusion largely prohibited Nokia from seeing the emerging smartphone trend and pushed it to hang on to feature phones market for too long. Eventually, when the rude awakening came, the confidence in the organization began to shatter, and the products started lagging behind in schedule.

While smartphones contributed a relatively smaller portion of the overall market, these high-end devices were a key growth driver. They got the most marketing dollars, got the most people talking, and pushed sales of mobile phone maker's other devices. A smartphone became a showcase for the company's entire product line. But just when the smartphone market got going, Nokia didn't have a worthy competitor to the iPhone, HTC Evo, Motorola Droid, or Samsung Galaxy.

Back in the mid-2000s, Nokia's segmentation strategy, which it had pioneered earlier in 1997, was seen as one of the answers to the company woes. In 2002, the Finnish wireless house had split its handset division into nine "mini-Nokias," each concentrating on a different market segment while sharing research, development, and manufacturing facilities. It wasn't just a matter of strategy; product segmentation was something embedded into Nokia's mobile DNA. It was Nokia who had brought the human aspects to the mobile technology—just like the early days of Apple Computer—by carefully segmenting the consumer market and designing phones for each segment. The Finnish wireless titan simultaneously launched both trendy and high-tech models, creating a challenge for rivals to keep up with such a pace.

But that was also partly the backdrop of how Nokia got used to the sleepy smartphone market. Ollila's matrix reorganization led to internal teams that were competing against each other in ways which the consensus management could not adequately resolve. The reorganization also established software development as a kind of factory production line from which product managers could pick and choose what they needed. Eventually, it became a key factor in Nokia's weakness in developing robust and powerful software products for smartphones.

Kallasvuo—after taking over the top job—insisted on remaining in all segments from 25-Euro basic phones to gem-studded luxury models. However, as part of his services drive, he bought the U.S. mapmaker Navteq Corp. for US$8.1 billion and vowed to boost the software organization with toolmaker Trolltech ASA and social networking start-ups. Kallasvuo drove the development of services such as music downloads and GPS navigation to increase the value of Nokia handsets and retain customers. In the end, these efforts did little to stop Apple and Android devices from taking market share from Nokia in the smartphone realm.

Under Kallasvuo, Nokia worried about hanging onto market share rather than creating innovative products that excited customers. Time and again, it vowed to fight back to this eminent smartphone leadership shift. But the company practically misread the way the mobile phone industry was merging with computing and social networking. Nokia was too slow to respond to the Apple iPhone, which redefined the smartphone market with its touchscreen, open web browser, and advanced application processor. The mindset at Nokia was that the company had tried touchscreens before, and consumers didn't like them. It couldn't be used with one hand. So there was nothing to fear. Consequently, as the iPhone sales took off, Nokia remained strangely detached.

Kallasvuo, an accountant cum corporate lawyer, had spent his formative years working in Nokia's finance department. He joined the company in 1980 as a lawyer and moved into finance in 1988. According to industry observers who closely followed Nokia, during his tenure at the top job, the Finnish firm hired a large number of executives who knew little about mobile technology, media, and design. Kallasvuo was also accused by many ex-Nokians of running an indecisive organization. Under him, for instance, an engineer could work on something for four years before a decision was made to halt it. Kallasvuo seemed to lack leadership to align and direct the company through these disruptive times.

So, in September 2010, when embattled Kallasvuo made way for Stephen Elop—head of Microsoft's Office software business unit—for the top job at the world's largest mobile phone maker, hardly anybody in the industry was taken by surprise. What really stunned the industry was what transpired soon afterward. The Microsoft veteran turned to his former employer for a turnaround plan and replaced Symbian as well as MeeGo with Windows Phone software as Nokia's principal smartphone platform. Elop also made radical changes in Nokia's management leadership and operational structure, and vowed to bring half of the top management from outside Finland. Nokia would now operate around two distinct business units: smartphones and mass-market basic handsets.

INNOVATOR'S DILEMMA

During the early 2010s, despite its entire residual market share, Nokia was starting to look like a has-been. Nokia's corporate culture failed to turn the "future gazing" into an agile strategy to advance its business; without visionary leadership and exceptional execution, good ideas remained just a series of disconnected dreams. In this section, the book presents two prominent case studies to show how Finland's most successful corporate leader blew up things on the chessboard of the technology landscape.

The iPhone, which boasted the most downloaded apps in games, became the first globally successful gaming phone platform to rival the Nintendo and PlayStation Portable. But the convergence endeavor in this specific market began many years ago with the first gaming-oriented smartphone: N-Gage. The device was Nokia's attempt at gaining market share from handheld gaming players like Game Boy Advance. In the early 2000s, gamers were increasingly carrying around both a cell phone and a Game Boy, the most popular handheld game system. Nokia spotted an opportunity in combining these two devices into a handier unit.

Apart from games, Nokia packed powerful functions into the system—multiplayer over Bluetooth or the Internet, MP3, audio and video playback and PDA-like capabilities. N-Gage created a plenty of buzz when launched in 2003. This combination of cell phone and gaming device was supposed to lure gamers away from their portable gizmos. But despite the large amount of attention gamers gave to N-Gage before it launched, the device earned scorn for its odd, curved design, and the fact that users had to hold the phone on its side to place a call. The speaker was on the side edge of the phone, resulting in many mocking it for side-talking.

The original N-Gage was heavily criticized for its clumsy design: to insert a game, users had to remove phone's plastic cover and access the battery compartment. Later versions fixed many of the problems with the original device, but the damage had been done. Gamers also blamed the lackluster performance on the poor selection of games compared to those available on Nintendo's Game Boy Advance. And yet

N-Gage cost more than twice as much. Nokia discontinued the phone line and related gaming services in 2010. In retrospect, it was an ambitious plan that was full of holes in the execution. In the end, the N-Gage account became a classical case study of Nokia's forward thinking that was mired with a flawed delivery.

The N-Gage debacle also showed that while Nokia was able to produce high-quality hardware, it didn't understand the crucial significance of apps and building an ecosystem around apps. The success of the iPhone in a domain where Nokia's pioneering effort failed yet affirmed that carefully crafted ecosystem played a crucial role in such innovative attempts. The biggest and most sustainable competitive advantage that Apple had over Nokia was the robust iOS services ecosystem. Nokia was now competing with companies that were born and bred in the digital era. But the Finnish mobile phone giant was simply not capable of matching the speed of innovation of Silicon Valley icons like Apple and Google because hardware was in its blood, not software. So it got bogged down in the alien details of building a sustainable software ecosystem and eventually wasted a lot of time.

Another glaring case study on how Nokia had been unable to turn its valuable research into products that consumers wanted: the company's flirt with near-field communication (NFC) technology. Nokia had been working on NFC since early 2000s and had spent millions on designing, developing, and testing NFC-based phones. Nokia made an early bid to bring NFC to mobile phones after the technology was born in early 2000s through a partnership between

Philips Electronics and Sony for catering to applications in the transport and convenience store segments. Near-field communication was an extension of radio frequency identification (RFID) technology that had been employed in the electronic tags used in road toll systems. In 2004, Nokia launched the NFC-enabled handset 5140 which came with tags that offered personal services shortcuts for browsing, text messaging, etc.

But the NFC test run didn't amount to much, and the 5140 handset remained largely undistinguished. Technology pundits hadn't forgotten the irony that wireless industry veteran Nokia envisioned the context-aware Internet for smartphones back in the mid-2000s. And that mobile newcomer Google actually executed that vision a couple of years later through its location and navigation apps for smartphones. Now once more, in a classical déjà vu, it was Google who stole the early march on mobile shopping when it launched the NFC-enabled Google Wallet in May 2012. The episode was another stark reminder of Nokia's inability to capitalize on its research to build products that could excite consumers. The NFC dud just added to a catalog of failures and missed opportunities, and symbolized what generally went wrong with Nokia in its quest for smartphone leadership.

A decade after Nokia surpassed Motorola to become the world's top handset maker, the Finnish company was troubled, confused, and conflicted. The mobile phone market changed tremendously during the first decade of the twenty-first century. The cell phone was where the consumer action was, but the ability to merely make a phone call was no longer a factor in selling these devices. People now wanted a single phone for their work and personal lives, and the new do-all gadgets did that bidding. Smartphones did everything from browsing the web to downloading and storing music and pictures. That's where troubles started for Nokia who had a history of innovation in the hardware space.

It was no longer about mobile phones; it was more about services and operating systems. In fact, the smartphone game wasn't even about operating systems anymore, but offering mobile users the features and functions they wanted in a carefully cultivated ecosystem. The battle of devices had turned into a war of ecosystems, and ecosystems included not only the hardware and software of the device, but developers, applications, commerce, advertising, search, social media, location-based services, unified communications and many other things. Apple and Google weren't winning market share with devices; they were taking Nokia's market share with an entire ecosystem. Nokia under the leadership of Stephen Elop quickly came to this realization that it was engaged in a platform war with Apple and Google. Now Elop had to decide quickly if Nokia wanted to build, catalyze, or join an ecosystem.

The ecosystem war started in 2007 when Apple launched the iPhone. While Google responded with the wildly successful Android, as chronicled in the preceding chapter, Nokia wasted a lot of time tinkering with the aging Symbian. What actually happened in the round two of the smartphone era was that, once microprocessors became more powerful, mobile devices began to transform into handheld computers. Consequently, most of the handset value shifted toward software and data services. That was the industry turf where the United States, especially Silicon Valley, possessed savvy business acumen. The companies like Apple and Google knew how to build overarching technology platforms. Silicon Valley boasted an unparalleled ecosystem of entrepreneurs, venture capitalists, and software developers who regularly spawned innovative products and services.

Apple's launch of the iPhone was the most instrumental episode in changing the mobile phone industry with thousands of handy applications. It provided clear evidence that services were now critical to smartphone success. In 2010, three years after Apple changed the mobile phone trajectory by making usability of the handset the new basis of competition, came the inevitable. The world saw the cell phone king Nokia struggling to develop a smartphone with the same mass appeal. Industry pundits blamed Nokia's weakness in smartphones for shaving about 70 percent of its market value during this period.

Olli-Pekka Kallasvuo—like his predecessor Jorma Ollila—was Nokia's CFO before he got promoted to the top job. Nokia, during his tenure, completely lost the plot and began to look like an NGO. Kallasvuo was widely criticized for his lack of connection with the industry and for his lack of charisma and substance to define the course of a big ship.

Image credit: *Taloussanomat*

Nokia managers loved "*Innovator's Dilemma,*" Clayton Christensen's management book about how successful companies fail to adapt to change. Ironically, the fate of N-Gage gaming handset became a stark reminder of the darker side of technology innovation.

Image source: *PC Magazine*

BLUEPRINT OF TURNAROUND

"Nokia: you make too many products. Focus on 3."

—A tweet from Twitter co-founder and CEO of Square Inc. Jack Dorsey on October 26, 2011

Text messaging had become so pervasive around the world during the first decade of the twenty-first century that mobile subscribers sent and received more text messages than they made phone calls. And the most compelling use of SMS-based business services was seen in the emerging markets like India and Kenya, where a growing number of consumers were using even the most basic cell phones to order food and flowers, do their banking, pay bills, make charitable donations, and purchase airline, bus, rail, or movie tickets. M-Pesa service, launched in Kenya in 2007, allowed people to send and receive cash through mobile phones, replacing banks in ordinary people's lives. The service became so popular that, by 2011, a quarter of Kenyan GDP passed through it. In India, Hindus even made advance bookings at temples via mobile phone to reserve the offering of prayers during busy holiday periods.

Nokia, which had largely missed the smartphone movement, but was still king of the hill in basic phones, saw an opportunity here. The European wireless titan began gearing the text-based applications toward countries like India that had a plethora of lower-cost phones and a preference for exchanging information through text messages instead of the mobile web. Short message service was low-cost, easy-to-use and didn't require a phone with service plan, and thus helped build services with a very low barrier to adoption. According to a *Forbes* report filed by Elizabeth Woyke, in November 2009, Nokia introduced a set of mobile programs called "Life Tools" which provided agricultural information and educational material to people in India's rural areas. Next up, in October 2010, the Finnish handset giant introduced two mobile applications that let Nokia phone users create chat groups and buy and sell products using text messages.

Unlike Life Tools, which supplied farmers with prices and other information, these applications targeted urban users and encouraged people to communicate with each other, not just consume content pushed to their phones. Take the example of We Meet service, a social-networking application that could be used by families, groups of friends, or small businesses. Users could create chat groups from their contact lists and communicate by sending text messages.

The application threaded the messages in chronological order, making it easy to follow the conversation. The effect was something like an instant messaging conversation, but at the fraction of the cost and on devices with no data plans.

The service allowed the phone's contact list to add notes about clients and accounts, and linked them tightly to the device's calendar. That way, a merchant could make a note about a future payment in the contact list and have it automatically saved to the phone's calendar as a reminder. The SMS-based service was designed to be location-aware. But instead of employing the pricey GPS technology, typically found in high-end phones, it tracked people's location via cellular towers. When people moved, the location updated. It wasn't exactly a digital map, but it served the same function.

Another service, called MoMart for mobile mart, comprised product listings delivered by text message. Interested buyers would subscribe to the service and specify the goods they wanted; the program would then push matches directly to their phones. Listings could be text-only or could include an image embedded into the message. Listings could also be targeted to particular areas using cell-tower location technology, enabling buyers and sellers to meet in person. Nokia managers liked to compare this digital marketplace to a Craigslist or eBay for India. These services might not look flashy, but they were still smart. Moreover, these applications were customer-driven, not just technology-driven. These simple but innovative services also craftily integrated two of the most promising mobile phone applications: text messaging and location.

Over the years, the SMS technology had evolved from simple text messaging to more sophisticated data services such as games and ringtone downloads. Back in the late 1990s, Nokia had introduced smart messaging protocol built on binary SMS rather than the standard text SMS, allowing mobile users to download ringtones and logos and to pay for them as part of their phone bill, generally by using a premium-rate number to access the service. One of the pioneering mobile phone successes had been these simple applications that allowed consumers to download ringtones and thereby personalize their phones.

Fast forward to early 2010s and Nokia was at crossroads. A low-price tag and longer battery life were no more the sole distinctions of Nokia's basic phones. It was imperative that the veteran wireless handset maker developed a more innovative and imaginative process to conceive good enough mobile products. Nokia boasted applications and services broadly available for specific regions and languages, particularly in Europe and emerging markets in Africa and Asia. Such applications and services could play a crucial role in holding onto existing customers and eventually shift them to smarter Nokia handsets.

Take mobile money as an example. The Finnish wireless concern could push its mobile payment and banking services across the installed base. Mobile money was born and raised in the emerging markets of South East Asia and Africa. Nokia didn't just work with top developers on a global level but had developer relations people in twenty one countries, from Beijing to the Middle East to Silicon Valley. It used these people to scout for well-performing regional apps as well as offer support to developers.

REINVENTING FEATURE PHONES

Much of this book has been dedicated to Nokia's smartphone problem, but it was also vital for the wireless firm to preserve its feature phone market. Nokia's mass-market phones—80 percent of its total units sold in 2012—were still a strategic part to its long-term survival. So reinventing feature phones, as well as reinvesting in emerging markets, was going to be a crucial part of Nokia's comeback plan. Nokia's new man at the helm had made a timely call to streamline the company's mobile handset operations around two distinct businesses: smartphones and mass-market mobile devices. Anything falling outside these two primary segments of mobile phones was simply taken out. Nokia, for instance, decided to unload its luxury handset brand Vertu to EQT VI in June 2012.

The risks to the Finnish handset maker remained stark even outside the smartphone realm because its low-end phone business was massively suffering from the overall market shift toward smartphones. Mobile device sales were rapidly shifting from feature phones to smartphones, so Nokia, like low-priced Android phones, started bringing sub-US$100 smartphone devices to the market. The company embarked on a strategy for making feature phones smarter by steadily bringing smartphone capabilities to other handset categories. If Nokia could hold on to that market with relatively budget-friendly Asha phones, and get those existing customers to switch to Windows Phone-based devices later on, it had a shot at staying in the game.

For a start, for mobile users without high-speed mobile data connection and Wi-Fi access, Nokia simplified Bluetooth connection through the Slam app, so that content could be transferred from one handset to another without any hassle. The system detected the nearest compatible device and paired automatically, allowing a mobile user to send content to the other device just by hitting the 'Send via Slam' option. Another cornerstone of Nokia's reinvigorated mobile phone strategy was connecting the next billion people to the Internet.

According to a 2012 survey from StatCounter, Nokia's Symbian platform, despite its cumbersome software legacy and steep market share loss, remained a leader in mobile web browsing. The Nokia Xpress browser incorporated the firm's cloud technology to reduce data consumption, speed load times, and shrink data downloads by up to 90 percent by compressing web pages before delivering them to mobile users. It was a crucial feature for mobile users who frequently used an older second-generation GSM connection or had a mobile phone contract with a low data allowance. The browser also supported more than 10,000 web apps—like Nokia Nearby—to give browser feature a greater functionality.

Nokia Nearby tapped the firm's mapping assets and cell-tower positioning technology to offer location-based services to phones that lacked GPS functionality. Nokia was also forming a partnership with Mozilla—the browser company—to integrate its location features in the upcoming Firefox mobile operating system. Nokia and its platform partner Microsoft were clearly lagging behind Apple and Google in the apps arena, but the "apps versus web" battle for smartphone riches provided Nokia and Microsoft with a window of opportunity. The web apps like Nokia Nearby could provide the Finnish phone maker with a level playing field; Nokia could use these web apps to facilitate moderately priced, Internet-ready models in emerging markets around the world.

Nokia's sales figures of the fourth quarter of 2012 suggested that the Asha smartphone line was gaining some traction in emerging markets. Symbian devices used to sell quite well in India, Eastern Europe, Africa and Latin America and these were the markets where Nokia likely had a strong opportunity with Asha phones. Nokia's Asha family of low-end, multitouch handsets was just barely classified as a smartphone. But calling these devices feature phones also seemed to be a misnomer. The Asha phones could send e-mails, link instantly to Facebook, and download free and paid games and apps from the Nokia Store. Even the camera integrated into the mobile phone was able to automatically resize pictures for easy sharing and posting.

But did the Asha multitouch range stand a chance against the Android onslaught? Elop was confident that the Asha handset could ward off the low-cost Android threat because it had a lower cost of ownership due to its browser's use of compression to cut down on data costs. Nokia also wanted to cross promote its homegrown S40 software-based Asha devices to Windows Phone developers and thus pursue a two-platform strategy in order to compete with the low-end Android devices. It's worthwhile to note that Nokia had enhanced Series 40 software to support touchscreen for its Asha line back in 2007, but the first full-touch Asha devices were shown to the public in 2012.

In 2012, Nokia's domination of the global market for cheaper, more basic mobile phones was inevitably coming under pressure from Chinese electronics manufacturers. There was clearly an opportunity in emerging markets that were up for grabs as Symbian sales declined. In emerging markets, such as China, where Nokia was still the most trusted brand, low-end local handset makers had started attracting mobile customers. Furthermore, the Chinese telecom manufacturers like Huawei and ZTE had started using their domestic success to expand globally, challenging Nokia's fragile foothold.

They were fast and cheap. A Nokia manager had acknowledged half-jokingly that Chinese electronic manufacturers were cranking out mobile devices faster than the Finnish phone makers' executives took time to polish their PowerPoint presentations. Inexpensive mobile phones based on free Android software and chips from Hsinchu, Taiwan–based supplier MediaTek were increasingly popular in developing countries. In 2008, MediaTek supplied complete reference designs for phone chipsets, which enabled manufacturers in mainland China to produce phones at an unbelievable pace. In 2012, by some accounts, this ecosystem comprising of MediaTek hardware and Android software produced more than one-third of the phones sold globally, mostly taking share from Nokia in emerging markets.

Nokia's failure was partly related to its inability to beat its competitors in the global feature phone market, where Nokia once dominated. Apparently, staying competitive in developing markets was proving to be tough for Nokia as cheaper Android devices flooded the market. The fact that the Chinese electronics stalwarts like Huawei and ZTE were in the Google camp, and they were rapidly eroding Nokia's baseline in emerging markets was a double-edged sword. They were flooding the market with devices that did all of the usual smartphone tricks. Then there was the dominant player and now the top phone maker Samsung battling Nokia at both high and low ends. Nokia was now fighting in the twilight world of feature phones and smartphones where smarts in the phone were a must anyway.

CLOSING THE CHASM

Nokia had been a highly efficient manufacturing and logistics operation capable of churning out a dozen handsets a second and selling them all over the world. Planning was long-term and new phones were developed by separate teams, sometimes competing with each other—the opposite of what was needed in software, where there was a premium on collaborating and doing things quickly. Take Apple, for instance, who had successfully demonstrated with its runaway success of the iPhone that the software interface was even more important than the actual product. Nokia clearly required being more than a hardware company to become a viable player in the post-iPhone smartphone world. The Finnish firm was without a tablet PC and a deeper focus on the creation of a broad software developer community.

Apparently, turning the European hardware-maker into a provider of software and services was no easy undertaking. At the same time, however, Nokia had to move fast if it wanted to have a chance to create a third platform for mobile software and services next to the iPhone and Android—hence the decision to ally with Microsoft rather than going it alone with a mystic combo of Symbian and MeeGo. Amid all the uncertainty, ceding control of the OS software experience to Microsoft seemed a logical and strategic move. It was essentially an ecosystem decision and was akin to starting over. By ceding control of the software experience to Microsoft, Nokia could stop trying to be a software company and focus on what it had always been good at: hardware.

There were a number of critical aspects of this deal that showed merits for Nokia. First, although Nokia would pay Microsoft an undisclosed royalty for licensing Windows Phone software, the strategic partnership would save Nokia a substantial reduction in terms of cost of developing and maintaining its own software ecosystem. Second, Nokia could tap Microsoft's software developer network, and together they could build a viable ecosystem to create a three-horse race that many in the wireless industry might welcome as an alternative to Android–iOS duopoly. Microsoft had a number of powerful service assets—from cloud services to Skype and Xbox to Outlook.

The cloud was now at the center of the mobile world, and that provided Nokia with a trump card in its long-term mobile renaissance. However, to make an impact in the cloud arena, it was imperative that Nokia released truly competitive devices not just with impressive hardware, but also with compelling software and services. For instance, synchronization among multiple device environments was the killer application in the connected era and Nokia didn't seem to be at par with Apple and Android at getting users seamlessly synched across multiple device platforms. It was apparently hard for Nokia, a company with telecom roots, to quickly line up cloud computing resources that matched the data center might of Amazon and Google. The iCloud challenges just showed the amount of stress and strain on Apple's computing resources.

Nokia's cloud conundrum could well be solved through the right partnership if not a refocused development. It was no-brainer that Microsoft's computing resources in general and its SkyDrive initiative in particular could significantly help Nokia bolster its cloud credentials. Microsoft SkyDrive integration added seamless cloud storage to mobile devices. The combined service assets of Microsoft and Nokia looked impressive and mutually beneficial amid the gathering cloud. Nokia's trusted software partner Microsoft had continued to push cloud-based services like Office 365 to compete with the likes of Google. One of the ingenious development concepts introduced with Windows 8 platform was compilation of applications in the cloud. Windows 8 would be a cloud-based experience through Windows Live SkyDrive feature that synched data, apps, and settings across all devices through a Microsoft account.

But arguably the most powerful asset the Redmond, Washington–based software maker had going forward was the range of service developers that created solutions for its broader Windows platform. That would offer a massive boost to the Windows Phone ecosystem, benefitting handset manufacturers. Next up, Nokia devices could take advantage of Microsoft's Office platform to try and appeal to IT departments as an alternative to both BlackBerry devices and bring-your-own-device (BYOD) trend which largely favored iPhone and Android gadgets. Microsoft's conviction in its Windows brand was very strong, and the firm was constantly trying to do more to leverage what it believed was the goodwill value of the Windows brand.

The Windows Phone user-interface also came across as a good complement to Nokia's design philosophy. Unlike Android, it didn't kowtow to Apple's styling paradigm. Instead, the user-interface was anchored in elegant graphics and simple and legible icons. Windows Phone 7 featured a more modern tile-based interface that Microsoft believed would charm customers. It was fast and easy to browse and navigate. Its next incarnation—Windows Phone 8 operating system—promised users to have the same experience on a PC, tablet or smartphone, and for many people on their televisions via their gaming consoles, and to do the same thing almost seamlessly from one screen to the next.

Although Microsoft was late to the game, Windows Phone was aimed at doing right what Google had been doing wrong. The crucial part of Microsoft's strategy was the quality control it imposed onto its hardware partners. Rather than coding an operating system and allowing mobile manufacturers to do whatever they want with it—like Google had been initially doing with Android—Microsoft required its hardware partners to meet rigid criteria in order to run Windows Phone. In its prior operating systems, even Microsoft allowed mobile carriers and manufacturers to determine the features they wanted on the phone; they would issue a list of specific instructions to OS makers like Microsoft. Consequently, mobile phones turned out to be overloaded with features and were unfriendly to users.

Windows Phone was badly trailing iOS and Android in the early going, but reviews of software had generally been good; as to the development of critical mass, the volume of handsets that Nokia delivered was expected to be a springboard for Windows Phone. The deal gave Microsoft access to one of the world's largest handset distribution networks. Microsoft had shipped more than 2 million licenses during the first quarter of the launch of Windows Phone 7 while Nokia had shipped 28.3 million handsets during the same period. If Nokia had chosen Android, it'd have probably been the end of the road for Microsoft's mobile ambitions. Now, at least on paper, Windows Phone had some momentum. Still, the alliance was more of a gamble, a last-ditch effort of sorts for both Microsoft and Nokia to gain a lasting foothold in the booming smartphone market.

Microsoft was desperately trying to make a comeback, and its proponents expected Microsoft to redeem itself with Windows Phone, as it did with Windows 7 after negative reception of Windows Vista. Microsoft had a history of eventually dominating the market in which it made a big push— until it set its eyes on the handheld and smartphone realm. Microsoft's hopes with Windows Phone aside, it had a staggering task ahead of it to catch up to and surpass Android and iPhone. As open-source codes emerged with graceful interfaces, created and managed by the likes of Google, it would become harder for mobile handset makers to justify paying royalties to Microsoft. It was all too well known that Sun Microsystems never really found a way to make money on Java amid fears that charging for technology would hinder its adoption.

That left Microsoft as the one scrambling for a piece of action in the mobile arena. But then Microsoft's multi-layer tie-up with Nokia came out of nowhere, bringing an abrupt about-face in the Redmond software giant's business model. It was a deal that was completely different from anything both companies had ever done before. First, it was a deal among equals. More startlingly, according to media reports, billions of dollars would funnel from Microsoft to Nokia most likely in the form of discounted software licenses, marketing cooperation and other non-cash benefits. Microsoft was reported to have paid Nokia US$250 million in platform support fees in the fourth quarter of 2011. After all, Nokia was keeping the Windows Phone platform in conversation.

The fight of mobile OS dominance was akin to the fight over next-generation gadget brains. Apple and Google captivated mobile phone users with sleek touchscreen software and app-phone bragging rights and, in just three short years, captured a large chunk of the market. But while the iPhones and Android handsets commanded much of the mindshare, Nokia continued to hold onto a significantly large market share. Nokia's global reach provided Windows Phone with a strong ally on the hardware side, and it was hoped that the union of Nokia and Microsoft would ultimately create a three-horse race.

CASE FOR THIRD ECOSYSTEM

Nokia and Microsoft were industry giants despite all their missteps and shortcomings. And they still possessed resources and expertise to fight with all their might to become a viable third platform. The mobile ecosystem was seen as large enough to accommodate and even benefit from a strong third player in the market. Consumers benefited from a strong Nokia because it kept the Android phone makers on their toes. Furthermore, mobile phone operators unhappy at being pushed around by Apple or Google could quietly favor the third mobile platform.

Most IT products were known to have standardized around one or two platforms. But the mobile industry did business on a different scale. The wireless business was unique in the sense that mobile operators had historically resisted monopolies or even near-monopolies. By 2010, there were other factors coming into play for the big business of mobile phone services. The revenue per user was starting to flat-line as mobile phone users spent more and more on games and apps that were not sold by wireless carriers. But the biggest fear that the mobile phone operators had was that they would be turned into dumb pipes without any control of the content and data flowing over their networks. The mobile operators' empire of controlling content, distribution, and policy had suffered the biggest blow in 2007 when the iPhone became the first mainstream device that didn't start up on a carrier-owned phone screen.

Apple transformed the wireless industry by enabling software distribution without going through operator stores, and thus created a brand new apps business by taking power away from mobile phone operators. Mobile phone companies didn't sell many apps because they were generally expensive and mediocre. It was a huge inflection point which symbolized mobile carriers being pushed to become the dumb pipes they had long dreaded. First, it was Apple and Google who changed the wireless business from being handset-focused to software-focused. Then, the mobile phone operators found themselves adjusting to a new power dynamic in which the big four of the tech establishment—Amazon, Apple, Facebook, and Google—were setting the trends and coming over the top to eat their margins and consumer mindshare.

Wireless industry partnerships had started to mature around the globe to stop these tech leviathans to get their foot in the door. The first major sign of the simmering tensions came in late 2011 when Verizon blocked the Google Wallet service from mobile devices operating on its network. The second largest U.S. wireless operator actually prevented Android handset makers from putting NFC devices onto mobile phones to ensure that Google's mobile shopping feature was a no-show. On the other hand, Microsoft, Nokia's software partner, agreed to play the mobile wallet game according to rules set by wireless operators in the Isis venture, later renamed Softcard to avoid association with the militant group in the Middle East. Microsoft apparently saw this as a window of opportunity to gain leverage in the war of ecosystems against mega competitors: Apple and Google.

There was another crucial dimension in the so-called war of mobile ecosystems. Nokia's most committed mobile partner Microsoft was striving to emulate Android's biggest advantage—enormous flexibility that allowed mobile operators and phone makers to customize whatever they saw fit—while evading its biggest drawback: fragmentation. The mobile operators were enormously important to Android's success. It was arguably not until Verizon's high-profile "Droid" branding campaign that Android started to take off. Google initially bent over backward to allow wireless operators and handset manufacturers to do whatever they wanted with the Android platform, even if it meant removing flagship features or denying users the ability to upgrade.

On the other hand, to keep fragmentation under control, Microsoft's position regarding liaison with wireless carriers wasn't as hostile as Apple's, but it wasn't far off. The Redmond, Washington–based software supplier had made some decisions that alienated mobile operators and manufacturers but that enriched user experience. With Apple being the most tightly controlled and, Android being fragmented and going all sorts of different directions, Microsoft in the middle could well be perceived as doing the balancing act.

Microsoft, while reworking its mobile strategy, had also borrowed a lot from Apple's playbook. The celebrity couple of Microsoft and Nokia that the industry had never expected to hook up virtually embodied an end-to-end ecosystem much like Apple. It was pretty ironic that Google seemed to emulate the same model through its acquisition of Motorola Mobility. This remarkable realignment in the war of ecosystems also showed that Elop's call for action regarding Microsoft deal was a far-sighted decision based on solid facts. Elop's strong conviction about ecosystem as a game-changer was evident from the fact that the Nokia leadership saw Google, not smartphone makers like HTC or Samsung, as its greatest competition.

An added advantage to Microsoft's bid for creating third ecosystem came with the landmark ruling in Apple's courtroom fight with Samsung. Apple had claimed that Samsung copied the design and functionality of the iPhone and asked for billions of dollars in damages. Apple's win in this legal battle set a precedent that left other smartphone makers vulnerable to Apple's patent claims. Apparently, the patent wars seemed to have put the otherwise unstoppable Android bandwagon on defense. So much so that Android flag bearers like Samsung and HTC had started looking for a safe alternative to Android while the smartphone market sorted out the impact of the Apple-Samsung legal spat. Both Samsung and HTC had also decided to launch Windows Phone-based handsets.

Last but not least, chipmakers, a crucial part of the smartphone value chain, wouldn't want the market to get polarized between Apple's iPhone and Samsung's Android handsets. Semiconductor firms were an important source of smartphone innovation, and they had crucial stakes in the mobile game. Remember, back in the 1990s, Nokia's liaison with TI for the development of chipsets was a key highlight of the mobile phone revolution. As a result, TI had just about sewn up the mobile handset silicon market. Then, during the mid-2000s, Steve Jobs gave the go-ahead for the iPhone project only when Apple engineers assured him that ARM-powered chips could handle the convergence of voice, data, music, and video. It was too hard to believe that hypercompetitive semiconductor industry would be content on shutting itself out of the smartphone design cycle and let just two handset heavyweights—Apple and Samsung—dominate the market.

Still, Microsoft and Nokia were fighting an uphill battle. Nokia's new software liaison was apparently a win for Microsoft, but it set Nokia on a highly risky and uncertain path with no guarantee of success. In fact, there was no silver bullet for either company given the strength of iPhone and Android. Nokia was walking a fine line in the twilight world of feature phones and smartphones. The early part of this chapter delved on Nokia's feature phone strategy. The next chapter will take a detailed look on how Nokia was feeding its Lumia strategy to make a comeback in the smartphone game.

The Finnish daily *Kauppalehti* claimed that Jorma Ollila's search for a replacement to Olli-Pekka Kallasvuo was a mere formality. He had already picked his protégé Anssi Vanjoki, who was getting ready to step in as the company's top manager. But then American investors forced Ollila to pick a CEO from overseas. Vanjoki quit almost immediately after it was announced that Canadian-born Stephen Elop would be the new chief executive officer.

Photo courtesy of *The Guardian*

The Lumia was the child of Nokia's marriage with Microsoft. Lumia devices—based on Microsoft's Windows Phone mobile operating system—won praise for their quality and style, but in the end, they were a bit late in the market. They were able to produce only mixed results.

Image: Nokia

6 THE BEAUTIFULLY DIFFERENT LUMIA

"It's an all-or-nothing bet. They have to be successful in the marketplace because there won't be anyone else to fall back on."

— Van Baker, an analyst at Gartner Inc.

The first Windows Phone device from Nokia—Lumia 800—was released in November 2011; however, the first device for the U.S. market—Lumia 900—didn't arrive until April 2012. The Lumia phones appeared reassuringly robust, and on the outset, seemed to rival the build quality of Apple products. Some reviewers even praised Lumia as one of the first true answers to the iPhone because it didn't pay homage to the iPhone stand-out design and was based around a unique experience. In the iPhone era, they said, Nokia was brave enough to forge its own path for differentiating Lumia from rest of the pack.

In smartphones, where call quality was often overlooked in favor of slick application processors and screen resolution, the Lumia handset offered a relatively crisp and loud audio. Next, after an ill-fated push into music services with Ovi Music some years ago, Nokia re-entered the entertainment business with the streaming Nokia Music feature. Another key highlight of the Lumia launch was Nokia doubling down on location and imaging technologies. The Lumia phone immediately got recognized for the photography innovation and mapping excellence. Stephen Elop gave a key priority to disruptions in areas like new materials, imaging, location-based services, and battery technology.

Nokia's smartphone struggles aside, its hardware design and feature set had always been a strength for the company. Nokia had some of the best technology in the business and a reputation for building high-quality hardware. Here, it'd be worthwhile to mention that it was Nokia who had made the external antenna disappear into the mobile phone back in the 1990s. Industrial design—the art of building appealing products with the required functionality at a target cost—was still Nokia's forte. Take Lumia's gorgeous handset screen that used super sensitive touch and worked even through gloves or mittens. The screen's bright colors and injection-molded polycarbonate body were meant to make it stand out in mobile phone stores where all handsets looked more or less the same.

Nokia made another claim to be the hardware vanguard by allowing mobile users to tailor an element of smartphone hardware to their individual needs. The embattled mobile phone company announced the release of 3D-printing development kit or 3DK to allow Lumia phone users utilize printable design files and instructions for making their own Nokia phone case—and customizing it however they wished.

Companies like Staples had already embraced the 3D print revolution, and the technology was being used in everything from toy manufacturing to cars. Now a firm of Nokia's stature was launching a 3D printing initiative to allow mobile phone users 3D-print their own custom phone cases, and that could pioneer customizability in smartphone hardware and potentially stimulate a whole new ecosystem. That also made Nokia unique among smartphone manufacturers. There were many 3D-printing schematics for iPhone cases that used Apple's specifications. However, they were not part of the core iPhone device.

PUREVIEW MAGIC

Windows Phone came with all the standard features that mobile users would expect in a smartphone: a reasonably fast browser, maps, e-mail, etc. However, the Lumia stood out well amongst the competition for its prowess in camera technology, which was the result of five to six years of work at Nokia labs. The Lumia handset was easily the best camera phone around, particularly for low-light conditions. Lumia's 8.7 megapixel camera performed very well, and one megapixel camera on the front provided a good image for video calling. Nokia's new phone was powered by an operating system that seemed to be better-suited to an era of large displays with very densely packed pixels. However, while the operating system created an environment that provided mobile phone design with hooks, the actual magic that made features like optical image stabilization work was hardware and software—which was done by phone manufacturer's engineers.

Nokia had made an investment in the PureView imaging technology so it could be seamlessly plugged into a mobile handset design. PureView was a pixel oversampling technique that reduced an image taken at full resolution into a lower resolution picture, subsequently facilitating higher definition, lighter sensitivity and lossless zoom. Nokia's Juha Alakarhu and Eero Salmelin came up with the idea in 2007 when they were on a business trip to Tokyo. They were both working on camera phones and were struggling with the challenge of using the zoom in these devices. Picture quality was simply not good enough, and part of the problem was the physical size of the imaging components. Everything in a mobile phone camera had to be small, which effectively translated into poor image quality.

Alakarhu and Salmelin came up with this radical idea of cramming a 40-megapixel image sensor into a mobile phone. They sketched the initial concept on a paper napkin. Eventually, Alakarhu and Salmelin developed the PureView technology along with some 400 engineers and technicians over a period of five years. PureView was a landmark imaging technology that went over all the approved limits. The effort culminated in the 808 handset, which boasted a 41 megapixel sensor: the highest and largest sensor in a camera phone at the time of its launch at the Mobile World Congress floor on February 27, 2012.

The 808 camera phone—offering more than five times the light-gathering capacity of image sensors in rival smartphones—stunned the industry with its easy-to-share images of exceptional quality. But there was a downside. The cost of using a larger size imaging chip measuring 13.3mm—when the rest of the industry had managed with image sensors no bigger than 5.7mm—became a luxury on a smartphone footprint. So the 41-megapixel image sensor had to be taken out of the early Lumia designs; though Nokia planned to incorporate 808-like image sensors in upcoming Lumia models. However, the early Lumia phones inherited the photo-rendering algorithms developed for the PureView technology.

There were two broad factors that defined how good pictures would be: hardware choices like image sensors and lenses that let in more light and software algorithms that rendered clear and bright images with pop.

The early Lumia phones came up as a disappointment compared with their illustrious camera phone predecessor, the 808 handset, but with 8.7 megapixel sensors, Lumia could still out-shoot most other smartphones, especially in low-light conditions. The wide f/2.0 aperture and floating-lens technology provided robust image stabilization and thus enhanced low-light pictures. The Lumia 920 phone borrowed PureView camera lens for the image-stabilization technology. Later, in 2012, Nokia pushed the technology envelope by packing a 41-megapixel camera into the Lumia 1020 handset and introduced 3X zoom feature.

The imaging guru Damian Dinning spearheaded Nokia's drive to catch up in camera smarts by making use of superior photo and audio features as a distinguishing mark. Dinning, a Kodak veteran, started working at Nokia in April 2004 and took the role of the company's lead program manager for the Imaging Experience division. He had worked closely on devices like the Nokia 808 PureView and Lumia 920. In November 2012, Dinning left Nokia to take a new position at Jaguar Land Rover, saying that he would be using his knowledge in smartphones to help automobile firm make advancements in the field of connected cars.

Another prominent highlight of Nokia's turnaround efforts came in 2012 when it acquired Swedish imaging software company Scalado to further boost camera-related aspects of its handsets. Scalado was founded in 2000 in the Swedish town of Lund with the mission to provide e-commerce sites with a simple way to view high-quality images of products over low bandwidth connections. Subsequently, Scalado changed its focus to the mobile market with a vision to redefine and improve the capture moments on handset screens. The company had worked with Nokia on imaging software for over a decade. It owned more than fifty patents and patent-pending imaging technologies at the time of acquisition. Eventually, using the Scalado technology, Nokia created the Imaging SDK and packaged it into the Lumia developer kit.

Nokia launched its first camera phone back in 2002 and became the world's biggest digital camera brand within a couple of years. The Finnish wireless titan sold more camera phones than film-based simple cameras sold by Kodak— and that made Nokia the biggest camera manufacturer in the world. The large handset makers like Ericsson, Motorola, and Nokia were not in the business of creating professional cameras, but with the advent of camera phones, they found themselves on the right side of the smartphone disruption. On the other hand, when the mass market took a radical shift away from standalone cameras, major photo industry players had to abandon the camera-related businesses and shift to something else like professional imaging, scientific instrumentation or photocopiers.

It was now Nokia preoccupied with Carl Zeiss optics, Xenon zooms, and raising bar of megapixels. The Finnish handset maker had even started offering camera phone tripod mounts. Carl Zeiss lenses, LED flashes, and physical shutter buttons distinguished Nokia camera phones long before Apple shipped the first iPhone. Nokia had long produced simply the best camera phones as high-end feature phones, and in a way, Nokia brought the awesome camera to the phone market a bit too early. But the market for mobile photography was still ripe for what Nokia had got in 2012 in the guise of Lumia. The Lumia handsets with robust imaging features were seen as having reinvigorated the camera phone market.

The difference between Nokia's relationship to Windows Phone and Samsung's to Android was that many of Nokia's feature hooks were promised to be integrated into the rest of the Windows Phone platform. Nokia promised consumers and app developers with three key differentiators. First and foremost,

Nokia introduced Imaging SDK with the Lumia handsets and offered developers the ability to build apps that took advantage of Nokia's unique camera technology. Second, Nokia provided location support with its HERE platform just like Google's Maps available on smartphones and tablets. Third, Nokia had its own music service for Lumia devices, Nokia Music, which allowed seamless downloads and purchase of music.

THE LUMIA CROSSROADS

Lumia was a proof that the partnership between Microsoft and Nokia made sense. But could Lumia save Nokia and Microsoft? Early sales had been fine, but not spectacular. While the Lumia phone was a good-looking piece of hardware, and Microsoft's tile-centric user interface had gotten good reviews, it was hard to see how that alone could be enough of a differentiator for both Nokia and Microsoft. The fact that Lumia handset sold just fine also challenged the paradox that all Nokia needed was one killer device to get back in the game.

It was apparently hard for Lumia to singlehandedly change either Nokia's or Microsoft's mobile market position because the modern smartphone market had turned into a two-horse race. It wouldn't be easy for the Finnish handset maker to get back the momentum it had lost to Apple, Google, and Samsung over the last few years. So Elop was content on calling Lumia a beachhead in the war of ecosystems.

The Lumia handset represented the culmination of Nokia's crucial transition from Symbian to Windows Phone software platform. However, by the time Lumia emerged on the mobile scene, most of the early smartphone users had already invested in either iPhone or Android apps, making it a barrier to switch. Therefore, despite some of the best technology that Lumia had displayed in the location and imaging domains, there wasn't enough reason for mobile consumers to switch from an iPhone or Android device. Perhaps a good audience to target for the Lumia devices was the first-time smartphone buyer.

Next, Nokia's smartphone start-over was up against a number of strategic and logistic hurdles. The Lumia handset had been well received but sales suffered as consumers held out for the next version of Microsoft's mobile OS software—Windows Phone 8—due in autumn 2012. Nokia launched its first Windows phone only to discover a few months later that its flagship Lumia devices wouldn't be able to run on the upcoming Windows Phone 8 software because they didn't have the hardware to support the newest features like NFC and Internet Explorer 10 web browser. The decision to make Windows 8 incompatible with previous versions made the early Lumia handsets look like stop-gap devices and made Nokia having to discount them rapidly.

The other elephant in the room was the app store. Nokia had succeeded in building a high-quality handset, and Microsoft was able to showcase its refreshing and likable operating system that ran Lumia. But the apps on Microsoft's Windows Phone platform were still in short supply. Even exclusive application partnerships with brands like ESPN and Sesame Street Workshop didn't help the Lumia phone to gain an edge because many developers still felt there was no compelling reason to create apps for Windows Phone devices. In 2012, although Microsoft claimed there were more than 170,000 Windows Phone apps, big developers like Instagram and Vine were reluctant to launch apps for Microsoft's mobile platform. Later in 2013, however, both Instagram and Vine announced Windows Phone apps at the annual Nokia World event in Dubai.

There were a small number of apps that a lot of smartphone users were interested in using, so both Microsoft and Nokia were targeting those applications and their developers. But Windows Phone lacked the long-tail apps strategy that made the iPhone so popular. One of the letdowns for a smartphone user was this perception that if he was looking for a specific app, chances were it would not be available on Lumia yet. App developers, mobile operators, and handset manufacturers tended to focus on platforms with a larger number of apps also because more apps meant more hardware choices and thus quicker product advances. The desktop industry had learned this lesson back in the 1990s.

The app phenomenon was often cited as a prime reason behind Apple's success. Developers have been a crucial resource for every major technology company's ecosystem. Apps ultimately determined what the company's product could do. Compelling apps also served as a major attraction for consumer sales. The BlackBerry OS and Hewlett-Packard's webOS primarily failed to mount a challenge to platforms like Android and iPhone because their advantage in usability, third-party software, and distribution was so paramount that developers would inevitably gravitate toward either iPhone or Android. An operating system hardly merited to anything if there were not enough developers seeding for the OS platform.

Apple, Google, and Microsoft were aggressively recruiting developers to make apps for their platforms because developers were foot soldiers in the war of ecosystems. Microsoft had already vowed to give developers a slightly bigger cut of app revenue than Apple and Google. Microsoft had also tried to encourage developers with various incentives, from free development tools to paying developers outright to build apps for its platform. The Redmond, Washington–based firm had surpassed 50,000 apps in less than a year after the launch of its Windows Phone platform. In retrospect, if Microsoft and Nokia kept launching devices like the Lumia, they could have managed to earn respect of both users and developers and thus get around the apps conundrum.

Finally, and probably most crucially, in the luridly colored Lumia hype, there was hidden an imminent danger for the Nokia comeback story. A predominant majority in Nokia's loyal customer base wasn't greatly amused to find the feature labyrinth on Windows Phone platform brought on by Microsoft's Young Turks. The complexity that came with adopting a new software platform seemed to be just the antithesis of the simplicity that was the hallmark of Nokia usability. Nokia's deep sense for the importance of basic form and usability was embodied in functions such as easy synching of mobile phone applications like contact details and business card sending, and that seemed being affected in the evolving Windows Phone world. For instance, in some cases, a Lumia handset user needed a Hotmail account to be able to use e-mail services.

In the transition from analog to digital mobile world, when branding was decisive, Nokia got the message better than anyone else did. The Finnish mobile stalwart was the first one to discover that handsets were too personal to trust to engineers alone. A cell phone was something that people kept close to their body, so it must have an emotional appeal. Just as much fashion and marketing went into a successful handset as good technology did. "Mobile phone is in a way an extension of a person's own personality," as a Nokia executive once put it.

However, Elop and team probably got so high on their Microsoft-centric agenda that they overlooked how critical it was to maintain Nokia's user experience strengths. Nokia needed to get that territory back in the unfolding Windows Phone chaos if it wanted to keep its loyal customer base intact. Elop's focus on new disruptions was spot-on, but it was also imperative that Nokia preserved the combination of elegant design and functionality that defined its mobile usability.

Clearly, the challenges for Nokia's special relationship with Microsoft went far beyond support from developers. Consumers in developing economies loved Nokia phones for their reliability, after-sales service, and thousands of informal service centers that knew how to fix Nokia hardware. In countries like India, for instance, there was a thriving resale market for Nokia phones. In India the word

"Nokia" had become a generic term—much like Kleenex and Xerox—and people were often heard saying, "Call me on my Nokia," even if it was a Samsung phone. However, in an abrupt about-face, mobile users in these emerging markets were reluctant to upgrade to Nokia smartphones running Windows Phone, because they generally found the software platform clunky and unintuitive and the fact that it offered fewer apps.

TOO LITTLE TOO LATE

In 2013, more than two years after Nokia's software marriage with Microsoft, the European mobile titan was far from finished playing out the Windows Phone strategy. The Lumia handsets were a world apart from Nokia's old Symbian handsets, so the expected turnaround was inevitably going to be steady. Moreover, there were still big gaps between Lumia and its competitors in terms of the functionality and usability of apps. That inevitably meant that there was a lot of work to be done on Lumia phones.

In the previous chapter, I explained why Nokia was betting on Microsoft's Windows Phone platform. The deal between Microsoft and Nokia was by far one of the biggest structural changes that had ever taken place in the mobile phone business. Nokia's ability to leverage its supply chain could bring Windows Phone to a bigger market and the volume of handsets that Nokia delivered could certainly be a springboard for the Microsoft ecosystem. Nokia also boasted a global reach and more boots on the ground than any other mobile player, and that could help Windows Phone to attract developers from around the world.

Windows Phone platform didn't gain traction despite the Nokia deal and BlackBerry's near collapse, but at least, Microsoft was fighting back in mobile. Microsoft had a history in being patient and persistent. For Nokia, there was also a silver lining in the fact that Microsoft had created successful products like Xbox 360 through a combination of force of will and deep pockets, and it was trying to do the same with Windows Phone. However, this time around, there were land mines everywhere for the agendas of both Microsoft and Nokia. All of the innovations on the part of Nokia and Microsoft did little to turn the heads of users who were still gravitating toward the iPhone and the increasingly popular Samsung Galaxy S franchise.

In 2012, Nokia sold less than 6 percent of all smartphones sold. Nokia's products had won praise for their quality, but they had arrived late. Sales of Lumia phones had been growing, but not fast enough to offset massive drops in other products. Furthermore, the early Nokia phones running Windows Phone software were expensive and failed to excite consumers. Nokia phones powered by Windows Phone operating system were neither affordable enough to appeal to cost-conscious mobile users generally opting for cheaper Android phones nor slick enough to win over buyers of high-end gadgets from Apple and Samsung. In fact, it wasn't until Nokia began expanding its portfolio to include more affordable Lumia phones that its market share position began to tick up. The model that helped Nokia to find some volume traction was the Lumia 520, which debuted in spring 2013.

The smartphone market continued changing at a relentless pace, and Nokia, a firm where workforce expected to stay for life, just couldn't respond fast enough. In the hindsight, the transition from Symbian to Windows Phone proved far more challenging than originally anticipated. The older-model phones were dumped, and newer phones with Microsoft software platform did not arrive on the market for months. In the meanwhile, Android bandwagon continued its relentless march across the globe. So the odds to reverse the fortunes of two laggards in a cutthroat smartphone market proved too big for the partnership between Microsoft and Nokia.

Nokia was now at a critical juncture in its historic transition, and all eyes were on the pathway for the Volvo of mobile phones to regain its former glory. In retrospect, Nokia seemed to have underestimated two things: the challenges of transition from Symbian to Windows Phone and the momentum behind Android platform. What probably damaged Nokia the most was the fact that it didn't have a striking low-end Windows Phone device that could compete with bare-bone Android handsets. And the fact that Asha line of entry-level smartphones proved to be lightweight against the onslaught of cheap Android phones. What next? Nokia management had an idea.

Juha Alakarhu and Eero Salmelin challenged conventional camera phone physics to make way for PureView imaging technology. Nokia's camera innovators played a crucial role in helping Lumia handsets accomplish an edge in the imaging technology.

Photo courtesy of Nokia

Lumia phones are seen here in the hands of Stephen Elop, but in a way, the opposite was also true. Elop's future was in the hands of Lumia phones. The beautifully different Lumia devices couldn't rescue Nokia, but they were able to save Elop's career and help him land a second chance at his former employer. So that he could carry on with his unfinished business of reinvigorating the Windows-based smartphones.

Image credit: *Engadget*

7 THE ELOP EFFECT

"It's often hard to see a challenger when you're dominant, but what happened with Android was faster than anything we've ever seen."

—Stephen Elop, the first non-Finn to run
Nokia in 145 years

Both Microsoft and Nokia knew that they were in this together. Apple's iOS was not an option for Nokia or any other Apple competitor. The obvious choice would have been to jump onto the Android bandwagon. Attracted by Google's free software, every other major smartphone maker except Apple and BlackBerry had already done so, and almost every large mobile operator was supporting Android as the only genuine Apple alternative. Back in 2011, Google was riding so high that it essentially refused to negotiate with Elop, offering no concessions to Nokia despite its global presence. He recalled Google acting like it had already won the smartphone battle.

Elop said it only made him more determined in his conviction that Apple and Android deserved some real competition. Nevertheless, his final assessment was that Nokia would not be able to differentiate enough on Android's platform. In retrospect, the financial struggles of Android phone makers such as HTC, Motorola and Sony showed that there was some merit to Elop's stand on the mobile platform owned by Google. In fact, no one except Google, Samsung and Microsoft actually made money off the sales of Android handsets. And Microsoft's revenues came through patent royalty payments.

Still, there were plenty of people suggesting that Android could refloat Nokia's sinking ship and that eventually Android would be the way to go for Nokia. Some hackers even tried to make it happen by getting Nokia's well-reviewed N9 handset run on Android Jellybean software. So industry watchers continued to question whether or not Nokia should have chosen Android instead of Windows Phone. If Nokia had done so, there's no doubt it would be selling more devices, though potentially at the cost of its soul. Android was made for everyone willing to drop the go-it-alone strategy that doomed Symbian. Android would have allowed Nokia to turn around a new handset in short order, but locked the mobile phone icon into an ecosystem where it would struggle to be a leader. Nokia would be subservient to Google in terms of the core software.

Nokia was, in fact, the swing factor in the post-iPhone era mobile game. The Finnish mobile phone titan could swing it to Android or swing the industry over to create a third ecosystem. Nokia chose to partner with Microsoft because it didn't want to become just another Android phone manufacturer. The partnership with Microsoft gave the Finnish handset maker a seemingly better shot at being a distinctive smartphone player rather than being yet another maker churning out Android phones. Furthermore, Google's platform partners like Samsung and HTC were already established hardware players who understood what was required to make the grade with an Android handset.

But pundits still questioned if Nokia would continue to hang its hat on Windows Phone if it became more and more apparent that it had no buzz or apps. Everything was resting on Nokia's gamble on Windows Phone platform, and if that didn't work, it could cause an existential crisis at the Finnish firm. Elop's vision was highly risky in the sense that Microsoft's track record in the mobile industry had been shaky. Microsoft wasn't exactly a neophyte in the mobile world. However, it had a troubled history that was mired with a long struggle without a single major breakthrough. In 2013, despite all the hype created through Microsoft's gigantic marketing machine, the truth was that Windows Phone wasn't seen at par with Android and iOS both from technical and market share standpoints. That had to change if Microsoft, as well as Nokia, wanted to stay in the mobile game.

Ironically, this time around, Nokia's set of problems with smartphones was not explained by a failure of execution. Now it was the Finnish phone maker's strategy that was becoming a problem. The book has made the case that Elop had some merit in making an ecosystem-based call in choosing Windows Phone to fight the duopoly of Apple and Android. However, in 2013, it was Windows Phone that was holding Nokia back. Windows Phone was still not a viable third platform, and Microsoft continued to lag behind in the pace of ecosystem development. In fact, Microsoft had many of the same problems that Nokia had in terms of innovation, especially in the smartphone business, and that didn't help Nokia. In effect, it was Nokia trying to stimulate the Windows Phone ecosystem through content deals to get the platform moving. The fact that Nokia was the biggest fish in the pond also meant that it would have a huge responsibility to grow that pond in conjunction with Microsoft.

What if Elop's high-stake bet on Windows Phone failed to bear fruit? Well, the Finnish handset maker could head to the Android camp anytime. In fact, in the early stages, the promise of differentiating smartphones on Windows Phone had actually been harder for Nokia than it had been for its rivals in successfully adopting Android. Microsoft's UI rules seemed to have made it hard for Nokia to innovate and differentiate while Samsung and HTC successfully built custom user interfaces and applications on top of Android. The fact that Elop was open to going Android was evident from his comments in an interview given to Spanish newspaper *El Pais* in January 2013. He said: "In the current ecosystem wars we are using Windows Phone as our weapon. But we are always thinking about what's coming next, what will be the role of HTML 5, Android... HTML5 could make the platform itself—being Android, Windows Phone or any other—irrelevant in the future, but it's still too soon [to tell]. Today we are committed and satisfied with Microsoft, but anything is possible."

Nokia was already walking a fine line in its courtship with Microsoft, which had shocked the tech world in summer 2012 by announcing that it planned to build its own tablet computer. That led to the buzz that Microsoft would eventually build its own smartphone as well. Nokia's Lumia phones didn't sell very well initially. In the meantime, Apple and Google only got stronger, making Nokia's situation even direr. Now questions began to emerge on what would happen to the special relationship between Nokia and Microsoft? Was Nokia proving to be the partner Microsoft expected? Would the release of the Surface tablet computer annoy OEM partners like Nokia? Would Microsoft compete directly with its most committed Windows Phone partner Nokia? Microsoft's history in the PC business showed that it wouldn't hesitate to deemphasize even a leading OEM partner if its business continued to decline.

Microsoft's tablet computer launch clearly marked a tension point in this perplexing partnership. One way to see this development was that Nokia had not taken a position on tablets yet. It's quite ironic that one of the first true mobile Internet devices (MIDs) to reach the market was Nokia's N Series Internet Tablet. But this gizmo—which debuted some two years before Intel's invention of the MID product category in 2007—had only met with limited success. Consequently, few portable device makers followed Nokia over the MID cliff. Fast forward to 2012, Nokia's weakness on the tablet front meant that Microsoft was just as far behind in tablets as it was in smartphones. And perhaps Redmond's computing leviathan didn't want to depend solely on Nokia for the tablet resurgence.

It's worth noting that before hooking up to Nokia, the Redmond, Washington–based PC software firm had been on its own and without much know-how of the cutting-edge mobile technology. But now Microsoft's biggest ally in the turnaround—Nokia—knew mobility all too well and could play an equalizer for Microsoft. Apparently, it was reassuring for Microsoft to have Nokia standing next to it in the fiercely competitive world of smartphones. Nokia had placed a strong bet on Windows Phone at a time when no other mobile phone maker would make more than a tepid commitment. In return, as mentioned earlier in the book, Nokia had won certain exclusivities that no other Windows Phone maker—not even Microsoft—had the right to duplicate. The contract granted Nokia the right to stuff almost any innovation it could muster into its Windows Phone handsets.

Nokia, for instance, had asked Microsoft to commit to using its HERE maps database as the foundation of the Windows Phone platform. In a way, Microsoft's Windows Phone software allowed Nokia to influence the direction of the newly developed platform. However, that also meant that the fates of Microsoft and Nokia were intertwined; Nokia's troubles had become Microsoft's troubles and vice versa. Amid all this uncertainty, Nokia's problem was simple: make Windows Phone a success or go out of business. Another intriguing part in this liaison was that Nokia had the option to exit the partnership with Microsoft by late 2014. And that was a frightening possibility for Microsoft.

THE ANDROID PUZZLE

In fall 2013, two news bytes—that came one after another— rocked the technology world. First, Microsoft announced it would buy Nokia's mobile device business in a US$7.2 billion deal. Second, trade press reported that Nokia had been working on an Android-powered handset ahead of the Microsoft deal. Nokia was making an Android phone while Microsoft was buying the company. Was it a mere coincidence? The truth of the matter was that the launch of the Android phone had underscored how badly Microsoft and Nokia each miscalculated the mobile market. For Nokia, its every misstep was becoming Samsung's gain. On the other hand, for Microsoft, it was the time to build on the Windows Phone story, not to start over with a blank sheet of paper: Android.

In the early phase of this partnership, Microsoft's Windows Phone platform didn't work on low-cost smartphones, and these devices eventually became high growth areas, especially in emerging markets. Microsoft and Nokia focused their Windows Phone devices to better compete with premium smartphones like Apple's iPhones or Samsung's Galaxy devices. Meanwhile, the missteps in the low-end smartphone market began costing Nokia huge amounts of lost volume. According to research firm IDC, in India, where Nokia phones held a big share of cellphone sales just a few years ago, Android powered 93 percent of smartphones shipped in 2013.

It's worthwhile to remember that Nokia had explored the Android possibility a few years ago, but the commoditization risk became the deal-breaker. Elop negotiated with Google about adopting Android before licensing Windows Phone and eventually decided that the cost—both in monetary terms and in terms of sharing any branding with Google— was too high. Nokia had also turned down an early chance to be Google's partner because its purchase of Navteq could have presented a conflict with Google Maps. Nokia wanted to replace Google Maps with its own location offering along with changes to Android's handling of e-mail, contacts, calendar, app store and over-the-air software management in an effort to stop value moving entirely to Google.

Fast forward to 2012, Nokia's Asha range of low-cost phones had come under intense pricing pressure when cheaper Android phones started to dominate emerging markets. That's when Nokia began working on a low-end Android mobile phone it internally called Normandy. The project was also known within the company as Asha on Linux (AoL). It aimed to repurpose the open-source version of Android into an entry-level smartphone as an alternative to Asha line, which was based on the aging Series 40 operating system. Normandy—just like what Amazon had done with its Kindle Fire tablet—used a forked variant of Android that was not aligned with Google's latest software version. The forked version of Android striped out the Google services just like Amazon had done for its Kindle devices. Amazon used the free Android Open Source Project or AOSP software on its Kindle Fire tablets and didn't pay Google for the platform and controlled the entire user experience.

Nokia made waves at the Mobile World Congress in Barcelona when it unveiled Nokia X phone on the show floor on February 24, 2014. The handset comprised of the Linux kernel; the open source pieces of Android; extensions that Nokia had made to that framework; and application programming interfaces (APIs) that Nokia built to replace Google Play store APIs. Nokia had built APIs for maps, push notifications and in-app billing that replaced Google's comparable APIs. That aimed to stunt Google's momentum because the mere Android OS wouldn't matter much. In fact, Microsoft would reap all of the ad and service revenues while keeping information from mobile device users out of Google's hands. Google gained no revenue and user data, and thus didn't benefit much.

So the Nokia X phone would in fact serve as a way to deliver Microsoft services such as Bing and Skype. At the same time, however, Nokia was wooing Android developers who wanted to build apps for users in developing markets. Nokia claimed to have tested more than 100,000 Android apps on the Nokia X platform and asserted that 75 percent of them were directly compatible and ready to be published on Nokia Store. Nokia managers also boasted that it would take Android developers less than eight hours to replace Google's services with Nokia's corresponding services. Furthermore, developers would be able to develop and distribute their apps in a single APK file format targeting multiple stores.

In a nutshell, the idea behind Nokia X phone was to pair the open-source Android operating system with Microsoft and Nokia services, and thus attain prices lower than the two companies had been able to hit with Windows Phone devices. Nokia managers believed that the company's

Android-based Nokia X phones could also become a feeder for mobile users who could eventually upgrade to more powerful and more expensive Lumia devices running Windows Phone. For Microsoft, however, the project risked consumer confusion and further entrenchment of an Android-first mentality among developers. Not surprisingly, therefore, Microsoft wasn't too happy with Nokia's decision to go Android.

Joe Belfiore, who ran Microsoft's Windows Phone division, echoed that sentiment at the same Mobile World Congress floor where Nokia X had been launched just recently. "There are some things they do that we are excited about and other things that we are not so excited about," Belfiore said while answering a question about Nokia X. However, in spring 2014, after taking over Nokia's mobile device business, Microsoft carefully reviewed pros and cons of Android-powered handsets. That was evident from the fact that Microsoft had continued the Android push when it launched Nokia X2 phone under its watch in June 2014. However, a month later, Microsoft discontinued Android-based Nokia X phone lineup as well as Asha handsets and other Series 40 feature phones.

MICROSOFT'S CASE FOR BUYING NOKIA

In 2012, Nokia's smartphone market share had plummeted across the globe, mostly due to cheap Android phones. The company's position had also become increasingly tenuous because Nokia had to do everything while working within the constraints of Microsoft's poor legacy in mobile and an operating system that didn't leave much to the mobile user imagination. Jorma Ollila later acknowledged that Nokia wasn't successful in using Microsoft's Windows Phone software to create an alternative to the two dominant companies in the field: Apple and Google. Nokia's Lumia smartphones had generated massive losses because of the low demand, and on top of that, its feature phone sales had started to collapse. That pushed the panic button at Nokia's power corridors.

In 2013, Nokia had just three percent of the global smartphone market, and its market cap was a fifth of what it was in 2007. According to market research firm IDC, during the second quarter of 2013, Microsoft's Windows Phone software ran on 3.7 percent of the smartphones shipped, compared with 13.2 percent for Apple and 79.3 percent for Android. Nokia phone sales were falling even faster, and by February 2013, the problems had become so acute that the Nokia board began to negotiate with Microsoft about selling the company's mobile phone operations.

Microsoft, on the other hand, was under enormous pressure to reinvent itself in a world where mobile devices were becoming the driving force in the technology world. Two powerful pillars of its business—Windows operating system software and the Office suite of applications—were tied too closely to the health of the PC market which had been suffering one of the most prolonged decline in tech business history. At the same time, some top managers in Microsoft seemed impressed from Apple's way of making products, bringing hardware and software under a single roof where they could be more elegantly woven together. Software had not only supplanted hardware, it also needed hardware as an ancillary business, they argued. Microsoft's unexpected launch of the Surface tablet computer clearly underscored that strategic thinking.

Finally, Microsoft increasingly felt threatened that Nokia might eventually stop making Windows Phones. So in February 2013, on the eve of the Mobile World Congress in Barcelona, Steve Ballmer reached out to Nokia chairman Risto Siilasmaa and initiated the talks about the purchase of most of the Nokia organization. However, the Microsoft board initially rejected the deal as too expensive and complex, arguing that the Redmond software house didn't need Nokia's handset division as well as its mapping unit. At that time, the board raised this fundamental question: should Microsoft be a software company or a hardware company too? According to media reports, Ballmer had even threatened to resign in case the deal didn't get trough. The argument was so loud that it could be heard outside of the closed doors of the conference room.

It was that particular moment in June 2013 when *The Wall Street Journal* reported that Microsoft and Nokia were in discussion for the sale of Nokia's mobile-phone business, but the talks fell apart over the price of the transaction. Subsequently, the handset unit-only deal was hammered out, and it included bringing Nokia CEO Elop—a former Ballmer lieutenant—back as head of the new devices unit. Then, in August 2013, shortly after the deal was finalized, Ballmer announced his resignation as Microsoft CEO. According to industry insiders, Microsoft board reluctantly agreed to the Nokia acquisition, but the blowup on this deal became the last straw in Ballmer's departure from Microsoft's corner office.

In September 2013, when Microsoft's US$7.2 billion purchase of Nokia's mobile phone business came on the heels of what appeared to be a failed acquisition in June that year, the deal didn't come as a huge surprise. Microsoft swallowing Nokia's mobile devices businesses wasn't a totally unexpected merger given the backdrop of two companies' three years of a rocky relationship. For Nokia, the deal was a capitulation to the harsh realities of its deteriorating position in a rapidly changing mobile phone landscape. Paradoxically, the news still came as a shock to many because Nokia as a smartphone brand was effectively dead. It was also painful for Nokia fans because it seemed as if Nokia had finally turned the corner with its Lumia smartphone business just months ago.

The deal brought five-year decline of Nokia's smartphone business and three years of its collaboration with Microsoft to an inevitable conclusion. Microsoft closed the acquisition of Nokia's devices and services business in April 2014 and renamed the operation as Microsoft Mobile Oy. The name—in which Oy was a Finnish word for companies—was due to a legal construct to facilitate the merger and would eventually be changed. That meant any future Windows Phone devices built by the new division of Microsoft would be Microsoft-branded.

A team in Salo, Finland would tackle high-end Lumia phones and another team in Tampere, Finland would design more affordable mobile devices. Stephen Elop went back to Microsoft and led an expanded devices team as the executive vice president of Devices & Services group. He would report directly to the new Microsoft CEO Satya Nadella. Elop had replaced Julie Larson-Green as the head of Microsoft's Devices and Studios business, which put him in charge of Xbox, Microsoft Surface, and Microsoft's game development efforts, in addition to the new mobile handset business. Larson-Green would now head the company's Applications and Services group, managing the look and feel of products like Bing, Office, and Skype.

Microsoft had paid about US$5 billion to buy Nokia's mobile phone business while it was shelling out another US$2.17 billion to license the struggling Finnish company's patent portfolio. Nokia would retain much of its patent portfolio and would grant Microsoft a ten-year license to its patents from the time of closing of the deal. Conversely, Microsoft granted Nokia reciprocal rights to use Microsoft patents in the HERE services. In addition, Microsoft would become a strategic licensee of the HERE platform and would separately pay Nokia for a four-year license. Microsoft was gaining access to over 30,000 patents from one of the two most valuable portfolios in the wireless industry. The other important wireless patent portfolio belonged to chipmaker Qualcomm.

The Redmond, Washington–based software giant had sealed the deal from a position of financial strength. Microsoft generated more than US$70 billion of annual revenue and the Nokia acquisition barely dented its US$77 billion cash stockpile. The price was too good to pass up for

Microsoft, which paid less for Nokia's smartphone business than the US$8.5 billion it had paid for the video communications service Skype back in 2011. It's worthwhile to mention here that the mobile devices operations that transferred to Microsoft had generated almost 50 percent of Nokia's net sales in 2012.

The "Nokia" brand would remain the property of the Finnish corporate icon; however, there were some time restrictions on the Finnish firm's use of the Nokia brand for smartphones. The "Nokia" name could not be licensed to another phone manufacturer for thirty months, and Nokia could use the brand on its own smartphones only after Dec 31, 2015. Microsoft had licensed the "Nokia" brand for use on mobile phones—not smartphones—for ten years, while the Lumia and Asha brands were transferred to Microsoft as part of the deal. Microsoft had bought the Asha brand as part of the deal rather than licensing it as the Redmond firm did with other Nokia brand feature phones. That gave Microsoft the freedom to kill the Asha brand, and it actually killed Asha phone line along with other Series 40 feature phones later in July 2014.

Evidently, it was difficult for Microsoft to get its software and services run on S40-based devices like Asha handsets. Elop also announced on Nokia's Conversations blog that Nokia as a brand would not be used for long for Microsoft smartphones and that it would be eventually replaced by a new smartphone brand. There were also indications that Microsoft could ditch the "Phone" part from its "Windows Phone" branding and would instead use just "Windows" for its mobile products. Meanwhile, the Nokia brand was very much alive for the new Nokia that was now focusing on mapping and mobile infrastructure businesses.

Approximately 32,000 Nokians transferred to Microsoft, including 4,700 people in Finland and 18,300 employees directly involved in manufacturing, assembly and packaging of products worldwide. That gave the acquisition team a gigantic task of integrating 32,000 Nokia employees with Microsoft's 100,000 employees. However, in July 2004, Microsoft announced plans to let go as many as 18,000 people, and the bulk of the layoffs, about 12,500 people, would come from the Nokia devices and services business that Microsoft had just acquired. It's worthwhile to note that a staggering 40,000 Nokia employees had already lost their jobs during Elop's tenure at the company.

Nokia, on the other hand, had hit the reboot button on its loss-making adventure with Microsoft and began to rebuild the company around its mobile infrastructure business. Apparently, it wasn't a happy ending, but the deal gave the ailing Finnish company a sliver of chance to regain some of its lost momentum. In fact, Nokia, now being a much smaller outfit, moved to a position of financial strength and aimed to rebuild the company for next big thing: broadband mobile networks. The European technology concern had effectively resolved the risk of commoditization; it was now a networks, patent-licensing, and navigation company. The new Nokia comprised of Nokia Networks, HERE Maps, and Nokia Technologies. These divisions had collectively generated around 50 percent of Nokia's net sales in 2012.

Nokia, as chronicled in the preceding chapters, still boasted valuable resources and intellectual property despite its legacy of squandered opportunities in the smartphone domain. Now that its torturous journey in the smartphone labyrinth had come to an end, nearly half of Nokia was still standing. The new Nokia had bought itself a future, and it was ready for the next big fight.

Many critics, especially ex-Nokians, blamed Stephen Elop to be a "Trojan horse" for Microsoft. The plot of "Elop conspiracy" went like this: When Elop hastened Symbian's decline in spring 2011, he knew that if Symbian sales collapsed, Nokia would start bleeding cash. And that's how he was able to sell Nokia's handset unit to Microsoft at a fire sale.

Image source: *Bloomberg Businessweek*

The change of guard in Keilaniemi, Espoo was heartening for many Nokians. On April 25, 2014, after the completion of Nokia's mobile phone unit acquisition by Microsoft, a crane truck took down Nokia's logo and replaced it with Microsoft's logo on this iconic building that previously served as the Nokia headquarters and had been famous as Nokia House.

THE NEW NOKIA

"If Rajeev gets it, it's a signal that Nokia becomes NSN."

— Richard Windsor, an independent analyst at Radio Free Mobile, speculating on Nokia's new chief after Stephen Elop moved back to Microsoft amid the sale of Nokia's mobile phone unit

In April 2013, Nokia Siemens Networks (NSN) closed the acquisition of the wireless infrastructure assets of Motorola Solutions for US$1.2 billion. The deal made the network equipment venture jointly owned by the Munich–based Siemens AG and Nokia Oyj the second largest telecom infrastructure provider worldwide after Ericsson and the third largest player in North America after Ericsson and Alcatel-Lucent. The NSN joint venture was announced back in 2006 when Nokia and Siemens combined their network hardware businesses to make a better impact in a highly competitive business. The purchase of Motorola's equipment gear unit was also an early indication that Nokia was taking its network infrastructure business far more seriously in order to counter its steady but massive decline in the mobile handset business.

In fact, Stephen Elop had hinted about the strategic importance of Nokia's infrastructure business soon after he took charge at Nokia. He had vowed to strengthen NSN's business to position it as a more independent entity. Eventually, Nokia's buyout of the NSN joint venture and sale of Nokia's mobile phone unit all but completed the transformation of Nokia into a network infrastructure-focused company. In September 2013, Nokia had sold off its handset business to Microsoft and became agnostic to whatever happened in the mobile phone segment. It was probably no coincidence that just before that, in July 2013, Nokia had paid Siemens US$2.21 billion and ended the tie-up with its German partner to fully own the mobile-equipment manufacturing operation.

Nokia renamed the new operation as Nokia Solutions and Networks or NSN and made it a wholly-owned subsidiary. Eventually, the company dumped the division's new name—Nokia Solutions and Networks, or NSN—and settled on Nokia Networks. Its headquarters remained in Espoo, Finland. The fact that Nokia had brought Rajeev Suri to head the remaining Finnish company further cemented the notion that the new Nokia saw its future in the wireless equipment business.

Suri had overseen the dismantling of the venture when Nokia bought out its German partner. He had also been credited with turning around the company's once-struggling alliance with Siemens. Suri joined the company in 1995, and between 1995 and 2007, he was a part of operations across Europe and Asia in business development and market strategy. He became head of NSN's Services Division in 2007 and became NSN's CEO later in 2009. Suri, like Microsoft's new chief Satya Nadella, was a graduate of the Manipal Institute of Technology, India. He had studied electronics and communication engineering at Manipal Institute, located on a hillside close to the Arabian Sea in the southern Indian state of Karnataka.

Initially, Nokia Siemens Networks (NSN) struggled to gain traction, bloated with high costs and too many staff in an industry dominated by the heavyweight Ericsson and cheaper network equipment from Chinese rivals Huawei Technologies and ZTE. The venture had been struggling to compete in the cutthroat equipment market till 2010. NSN, like most of its competitors, was operating in both wired and wireless infrastructure businesses. But then it made the right call to make a shift from a full-services supplier to a specialist vendor and began ridding itself of non-core assets like services and fiber-optic business. In 2011, NSN launched a restructuring in which it cut 17,000 jobs and reinforced its focus to mobile broadband and related services. The focus on mobile broadband a.k.a. Long Term Evolution (LTE) equipment worked quite nicely for NSN.

In 2012, NSN's profitability improved significantly due to the growing demand of LTE-based mobile broadband infrastructure, notably in the United States, where mobile carriers were investing heavily in turbo-fast 4G networks. Nokia's often ignored infrastructure joint venture with Siemens now began to look upbeat amid impending growth in the LTE-driven wireless infrastructure business. And it became a source of much-needed cash for Nokia. In 2012, the wireless infrastructure supplier moved from fourth to second place after Ericsson in the LTE equipment market ranking. Same year, ABI Research ranked it the world's leading vendor in LTE contracts, intellectual property, and most notably, progress in small cells. Now NSN was trying to cultivate a burgeoning reputation of being on the cutting edge of next-generation 4G wireless networks.

In the hindsight, it wasn't in Nokia's larger interest to put all its eggs in one basket: mobile phone business. A concurrent grasp of wireless equipment and handset businesses had long been a panacea in the wireless telecommunications world and companies like Ericsson and Motorola had worked very hard on this premise without much success. The revival of network infrastructure business was surely good news for the Finnish multinational corporation. Moving forward, Nokia's wireless infrastructure operation, along with its location and patent businesses, could act as a strong deterrent to the challenges that Nokia faced in both feature phone and smartphone segments. That powerful antidote was hardly mentioned in the trade press amid torrid headlines about Nokia's smartphone debacle.

Now the Nokia reinvention episode was a classical déjà vu of what had previously happened to its Swedish neighbor Ericsson. The Stockholm–based wireless titan stuck to its mobile phone business as long as it could. Then, in 2001, Ericsson formed a handset venture with Sony in a hope to borrow consumer smarts from the Japanese electronics giant. However, in October 2011, Ericsson threw the towel and unloaded its stake in the struggling mobile phone venture to focus on serving wireless operators. Apparently, soon after Sony took over its mobile phone joint venture with Ericsson for US$1.5 billion, the sole focus on serving mobile operators came as a welcome relief for Ericsson. In retrospect, the consumer-focused mobile phone business ran contrary to Ericsson's engineering-heavy, business-to-business focus.

The tough call had also proved a blessing because mobile carriers were about to make a massive bid on LTE networks to satisfy the data-hungry smartphone users. Ericsson quickly cemented its position as the top wireless infrastructure supplier. Now Nokia was just aiming to do the same. In fact, the next incarnation of Nokia was starting to bring a somewhat similar scene. In August 2014, Suri told *The Economic Times* of India that it was the strongest financial position that Nokia had in five years with an 11.4 percent operating margin. He noted that if the devices business was included, that would bring Nokia down to zero percent, and no one would want to work for a zero percent margin company.

PATENTS TREASURE

The infrastructure unit, which supplied radio base stations and other wireless hardware equipment to mobile operators, generated almost 90 percent of the company's annual revenue when the handset deal was closed. Still, the new Nokia was keen on its other two businesses. The larger one was HERE, its highly acclaimed maps division, which had most of the market for navigation systems built into cars. The smaller one—Nokia Technologies—had the job of licensing Nokia's thousands of patents. Chairman Risto Siilasmaa had called Nokia Technologies the company's innovation engine.

The outcome of the courtroom battle between Apple and Samsung in summer 2012 was a verdict that found that Samsung had infringed on several Apple patents. The well-documented fight between Apple and Samsung over patented technologies showed just how important intellectual property rights were going to be in the fight for the smartphone gold. That legendary fight between Apple and Samsung was also a stark reminder of how far a strong patent portfolio could go and how this could be monetized in various ways. The crucial importance of patents in the new mobile order was also evident from the fact that around 80 percent of Android phones were covered by Microsoft's patent royalty agreements. According to some estimates, in 2013, Microsoft collected as much as US$8 per Android device, leading to an estimated US$4.3 billion in Android royalties.

Nokia's patents treasure chest was a hidden gem in the Finnish company's next incarnation. Nokia was now practically in a licensing business with traditional economics: it was getting paid. While Microsoft had picked up around 8,500 design patents, the bulk of the patent portfolio remained with Nokia after the purchase was complete.

Nokia had quietly built up a very strong and relatively young intellectual property portfolio and owned almost 16,000 patents around telecom in the United States alone and another 20,000 patents outside of the United States. A 2011 survey showed that Nokia was the largest patent holder for essential technologies relating to LTE networks.

Patents and intellectual property made up one of the three business units of the new Nokia. The Finnish firm made money through its patents on every iPhone sold. Motorola Mobility—which Google sold to the Chinese consumer electronics maker Lenovo—also paid fees to Nokia to license wireless patents. Next up, one of the salient highlights of the sale of Nokia's mobile devices business was that Microsoft would pay Nokia for a four-year license of the HERE services, bringing the new company more revenue and stability than it had previously. It's worthwhile to note that Nokia's mapping unit had generated 7 percent, or US$1.2 billion, of the company's total revenue in 2013. Nokia had continued to invest in the mapping operation, which comprised of 6,000 employees, or around 11 percent of Nokia's remaining workforce of 55,000 staff members.

The Navteq part was now clearly one of the Finnish company's crown jewels. The geographical information system company prospered primarily because of its automobile industry partnerships. Its digital location system dominated automobile mapping services with more than 80 percent of the global market for built-in car navigation systems, a segment in which Google and Apple were scrambling to catch up. Navteq provided navigation apps to a lot of cars made by Volkswagen, Mercedes, Hyundai, and Toyota. Its data was present on four out of five in-car navigation systems globally, and it was a constant push by the automotive partners that had helped the digital mapping operation get as good as it was.

The more consumers used a platform, the better it got. The scale was imperative for a location platform to attain the highest quality. Besides automobile customers, Nokia licensed its HERE software to a number of companies including Microsoft for its Bing search engine, Amazon for the Kindle Fire tablet, and Yahoo for its Flickr photo service. FedEx and Oracle were also embedding HERE maps in their technology systems to direct deliveries worldwide and to manage their global supply chains, respectively. The deal with Oracle—the world's third-biggest software firm with a particular expertise in database management—was surely a reason to cheer for Nokians. Another major brand win was courier services giant FedEx that used HERE mapping information to manage its fleet of delivery trucks worldwide.

Nokia had announced a similar deal with online coupon and e-commerce firm Groupon. Then, there were mobile operators such as Deutsche Telekom, who were building location services that used digital maps and allowed mobile users to keep track of friends and family members through smartphone apps. Next, Nokia was collaborating with retailers, banks and tech companies to tailor its real-time mapping information—including foot and vehicle traffic data—to help them determine where to place stores, billboards, kiosks and ATMs.

Initially, Nokia had been focusing on the power and thoroughness of its mapping database—which had information on nearly two hundred countries—in an effort to distinguish its Lumia phones from the competition. However, over the years, the Scandinavian company's ability to furnish computational mapping data onto portable devices made it a Google-like platform player and a horizontal provider of location-based services. The new Nokia's goal with its mapping service HERE was simple but ambitious: it aimed to build the world's most detailed and up-to-date digital maps. As more companies connected their products to the Internet, mapping services would become increasingly important to more businesses, ranging from transportation to shipping to retailing.

THE REMAINS OF NAVTEQ

In 2007, soon after Olli-Pekka Kallasvuo took charge of Nokia as the new CEO, he embarked on an ambitious technology acquisition. Nokia bought the Chicago–based digital map company Navteq Corp. for hefty US$8.1 billion to bring navigation out of the car and deliver it to pedestrians. Nokia coined the buzzword "context-aware Internet" while asserting that it would reshape the Internet. To accomplish that Internet panacea, the Finnish mobile giant was pinning its hopes on operator-independent, cross-platform phones conceived through development of new software and services. The company claimed that Map 2.0 would enable context-aware Internet by combining multimedia features with the freewheeling Internet and assisted-GPS technology.

Nokia engineers asserted that by adding context—such as time, place, and people—to the Internet, the mobile web experience would become something entirely different. Once the context was added to the network, they contended, the Internet experience would become more mobile, contextual, and personal than on the desktop. The building blocks necessary to make this happen included GPS, broadband wireless access, a back-end service, and enough processing power and memory residing on mobile phones. Here is one scenario depicting how it would actually work: a user takes pictures with a camera phone, and the GPS coordinates are simultaneously stored in the metadata file; Bluetooth could sniff around and discover who is around him or her. Location, therefore, would no longer be an application; it would become the core fabric of the mobile Internet.

Nokia managers loved to play up fascinating new scenarios at technology press events. Their hyperbole was reminiscent to the early days of the mobile commerce talk that was stimulated by the arrival of WAP-based mobile phones back in early 2000s. At that time, marketing dream weavers conjured up whiz-bang scenarios in which mobile-phone users would resort to all kinds of amazing adventures. One might have heard this: walking down the street, a user approaches a Starbucks coffeehouse and his or her mobile phone starts ringing; on the handset screen pops up a coupon for a $1 latte. Or this: A user strides into a department store and slips into that perfect pair of jeans. A bit pricey! No sweat for his or her mobile phone! The user punches the barcode of the jeans into his or her handset and receives 20 percent discount from an online retailer.

However, these marketing potions carried a fundamental flaw: the key building blocks to make mobile commerce a commercial reality were not ready yet. What Nokia did in 2007 was recycle this notion by combining two chic technologies of the time—the mobile web and GPS-based location—and started spreading the context-aware Internet gospel. However, after three years and some failed projects, there was little evidence of any tangible payback to Nokia's foray into location-centric premium phones. Location-based services got off to a bumpy start in the early and mid-2000s also because mobile industry failed to integrate digital maps into appealing packages. Location-based apps—to summon a taxi, for instance—needed maps inside them.

Fast forward to 2010 and it was a very different story in the post-iPhone arena. Hundreds of millions of smartphone owners were using mapping apps like free turn-by-turn navigation; location was also bubbling away as a critical feature in social networks like Facebook, Foursquare, and Twitter. Maps and location became so critical on mobile devices that it would be considered crazy to have a smartphone that didn't include GPS and software providing location-based services. So much so that Stephen Elop would describe location and mapping as the third dimension of the search, representing the "where" element, with "what" and "who" being the first two dimensions.

Location and mapping were now a big deal. There was a big story in the trade press when the iPhone whiz Apple got the mapping services wrong. One of the most unpopular changes that Apple carried out in the iPhone 5 platform was to replace Google's Maps software with its own mapping app, a move that eventually infuriated many iPhone users. There were media reports that Google wanted to pummel iPhone users of Maps with ads, and that was apparently unacceptable to Apple. So Apple decided to put its own maps into the iPhone 5 rather than be beholden to its archrival. However, Apple couldn't pull it off on its own. The new maps were built into the iPhone 5 handset that went on sale in September 2012. When consumers used Apple's new mobile maps, they found nonsensical routes and misplaced landmarks. While Apple's location service was widely criticized for being buggy and incomplete, the controversy also highlighted the importance of quality mapping technology to mobile users.

Meanwhile, Nokia began to double down on its mapping strength, positioning itself not only against Apple, who was having problems with its maps business, but also Google, who was pumping even more investment into its successful mapping services. It's ironic that Nokia's acquisition of mapping platform Navteq in 2007 was widely criticized for the high price tag. By 2012, however, Nokia's bet on location and mapping had turned into a steady progression of good news. Navteq's experience in the mapping industry dated back nearly thirty years, and those three decades of digital cartography work had helped the Chicago–based company build a thoroughly comprehensive set of mapping data.

The Navteq acquisition also brought a much-needed boost to Nokia's bottom line. First and foremost, comprehensive navigation services were baked right into Windows Phone platform. In the war of ecosystems, by elevating Windows Phone platform as a whole, Nokia was helping to raise its own business case as well. With global data from Navteq, Microsoft was also able to enhance the mapping service of its Bing search engine. Bing had superior mapping details in the United States but not abroad; now data from Navteq could put Bing on a par with Google Maps around the world. Nokia Maps were regularly receiving excellent reviews when compared to other mapping solutions, including Google Maps, and some industry observers believed that Nokia's maps could even outstrip Google's.

Google's maps were good, thanks to years of work, massive computing resources, and thousands of people hand-correcting map data. Google had spent literally tens of thousands of person-hours creating its maps. The common perception was that no other company could beat Google at this game. As it turned out, even the reliance that Google Maps had on Navteq data was very high, and that was one reason the Navteq team invested heavily in augmenting the data with its own data set. Navteq offered turn-by-turn driving instructions in fifty languages in around hundred countries. According to 2012 figures, Google's Navigation service covered a mere twenty eight countries and one language: English.

Nokia was now the world's largest mapmaker. It had started building out Navteq True, a three-dimensional scanned database of positional information that could eventually make traditional, two-dimensional maps obsolete. The product was going to be a digital representation of the real world. Nokia was also going to acquire the imaging company Earthmine for its 3D street-level imagery. The Berkeley, California–based firm specialized in three-dimensional maps showing street views; it brought a comprehensive solution for collecting, processing, managing, and hosting three-dimensional street-level imagery complete with camera-centric mobile software. It was a clear signal of Nokia's ambition to go head-to-head with Google's Street View service while also beefing up three-dimensional data for its location apps.

Then there was LiveSight, a bundle of technologies that helped location apps detect buildings that were being viewed through a smartphone's camera. The augmented reality-enabled intuitive mechanism filtered points of interest to only show those in line of sight and froze the camera frame to inspect the location without having to hold the camera pointed at the target. The LiveSight 3D mapping technology was first being employed in Nokia's City Lens augmented reality app that enabled users to point the camera at real-world objects and see data overlaid on top of an image on the mobile screen. Pointing the camera at a restaurant, for instance, pulled up online reviews for it. That way the City Lens app put virtual reality to good use by superimposing points of interest on mobile users' surroundings.

MAPPING OUT HERE

The next major expansion in the remaking of Nokia as a navigation expert came with the announcement of the mapping and location intelligence business called HERE. Nokia had built the business primarily through acquisitions, starting with its buyout of Berlin–based gate5 AG in 2006, followed by the acquisition of Chicago–based Navteq for US$8.1 billion in 2007, and 3D mapping technology expert Earthmine in 2012. According to Nokia, HERE was the "world's first cloud-optimized version of mapping technology that delivered location platform, location content and location apps across any screen and any operating system."

HERE offloaded many tasks to the cloud, so that maps could load using less bandwidth and could be downloaded to a smartphone without taking up much storage space. However, the early versions of HERE had their own quirks. The look of the maps wasn't crisp and clean, and that gave the service an unfinished look. The single biggest problem was its bugginess: the app was mired with incomplete transit directions and other missing data. HERE maps provided satisfactory driving, walking, and public transit directions, but the app needed specific details from users. Nevertheless, Nokia demonstrated that it had the data and content to power great mapping experiences. Over time, HERE continued making a steady progress.

HERE included directions, 3D city views, live traffic data and the ability to create and save a collection of destinations. Nokia's rebranded mapping service was initially available through a web browser and as an iPhone app. Eventually, it became available on Android and Mozilla's Firefox operating system. Then, in 2014, Nokia signed a licensing agreement with Samsung, which brought Nokia's HERE maps to Samsung's Android-based Galaxy line of smartphones.

There were clear indications that Nokia management highly valued the company's location services division. The Espoo,

Finland, company hoped to get smartphone users hooked onto its mapping technology regardless of the mobile platform. Besides Windows Phone, Nokia wanted to see HERE a hit across different devices and software platforms—including iPhone and Android. So that, ultimately, the open digital mapping platform called Nokia Maps could seamlessly extend support across multiple mobile ecosystems. The forgotten mapmaker Navteq had now become a silver lining that helped Nokia to come back to the tech limelight. Nokia's advances in location and mapping services were also a reminder that there was value within Nokia beyond its handset manufacturing business.

According to industry estimates, HERE contributed less than 4 percent of the company's valuation but its potential for future growth was immense considering the rising demand and growing penetration of intelligent location and mapping services, especially in automobiles, smartphones and connected wearable devices. In summer 2014, Nokia had launched a US$100 million Connected Car Fund to identify and invest in companies that could help grow HERE's location and mapping ecosystem in the automotive space. Moreover, Nokia had made a couple of interesting deals to make the HERE unit more personalized and intuitive and boost the long-term potential of its mapping business.

First, Nokia acquired artificial intelligence firm Desti, a virtual travel agent that mined data from place descriptions and reviews and then used that data to generate more personalized and refined recommendations for sightseers and business travelers. The Seattle–based Medio Systems was the second company Nokia's HERE division bought to help it create more personalized maps that did more than just give directions. HERE navigation platform aimed to create highly personalized maps that anticipated a user's destinations and desires and Medio could help in providing that context. Medio started out as a mobile search engine in 2006, but it was forced to abandon the idea as Google came to dominate mobile search just like the desktop online search. Eventually, Medio put its contextual search algorithms to work to create predictive analytics models.

Mapping, however, was an expensive business. Nokia itself claimed that its mapping products were updated 2.7 million times a day. So it was still a question mark if Nokia had the deep pockets required to keep pace in mapping. Industry observers widely believed that Nokia could either decide to sell or spin off the division, so the company could focus on its core mobile infrastructure business. They argued that HERE might be more valuable to someone else than to Nokia. It could well be Apple who had tried to break into the global mapping business with little success. Or it could be hyper-ambitious Samsung whose stakes in the mobile game were increasing by the day.

Microsoft was apparently the usual suspect. The tech titan from Redmond had fought hard to buy HERE as part of Nokia's handset business sale, but the two companies couldn't agree on a price. Mobile phone and tablet makers mostly relied on Google for mapping services. However, as smartphones became more sophisticated, handset makers might want to take more control over this vital part of user interaction and consider having their own alternative to Google Maps.

Nokia was transforming HERE business into a location cloud. So that, beyond smartphones, it could serve areas such as low-power connected sensors, distributed sensing systems and intelligent interplay between various types of radio technologies. Nokia could either endeavor to turn the HERE unit into a big business in the imminent location era or balloon its worth and sell it at a good price. In any case— the making of a flourishing location technology business or a multibillion-dollar takeover target—Navteq's efforts to digitally map the entire world could prove to be Nokia's hidden gem.

In summer 2013, prior to selling its mobile handset business to Microsoft, Nokia bought out Siemens' 50 percent stake to take complete ownership of Nokia Siemens Networks (NSN). The venture—first renamed as Nokia Solutions and Networks or NSN and later settled on the Nokia Networks title—completed the Finnish company's shift toward the mobile infrastructure market.

Photo credit: *Reuters*

The "Indian Diaspora" had reached Helsinki, the capital city located on the shore of the Gulf of Finland, an arm of the Baltic Sea. The fact that infrastructure man Rajeev Suri was going to lead the new Nokia clearly underscored the premise that the Finnish firm saw its future in the mobile networks business.

Image source: *Helsingin Sanomat*

9 NOKIA'S PREDICAMENT

"We were not successful in using Microsoft's operating system to create competitive products, or an alternative to the two dominant companies in the field."

— Jorma Ollila, in a 2013 interview with the Finnish newspaper *Helsingin Sanomat*

Nokia's predicament was akin to the dark days at Apple during the mid-1990s when the personal computer pioneer was at the brink of bankruptcy. Stephen Elop had inherited a massively complacent organization that in many ways resembled Apple of the days before Steve Jobs's return in 1997. The irony was that the world's largest cellphone maker for more than a decade was looking to replicate Apple's success with consumers when it sought a new chief executive officer in 2010. However, industry pundits said that Elop was too late, and his attempted rescue was most likely doomed. They saw a lot of downside for the mobile phone maker and maintained that Nokia—a case study in corporate turnarounds under Jorma Ollila in the 1990s—couldn't be turned around this time.

Despite Microsoft's valiant efforts, Windows Phone was off to a slow start, so the risks to Nokia remained stark. Further into the turmoil, the changeover from Symbian handsets to ones produced by the new Microsoft alliance represented the point of maximum peril and uncertainty for Nokia. Not surprisingly, therefore, the transition years of 2011 and 2012 embodied the gravity of Nokia's predicament. The shift from Symbian to Windows Phone required a significant capital investment, but the nosedive in Symbian-based smart-phone sales had crushed Nokia's revenue base, further constraining its cash position. The BlackBerry case study showed that OS transitions could prove far messier than originally perceived. The switch from BlackBerry 6 to QNX operating system had turned out to be a massive strategy shift that shook the Waterloo, Canada–based smartphone maker.

Then, in summer 2012, came the inevitable: according to an IDC survey on smartphone sale, the once-dominant Nokia had fallen out of the top five smartphone sellers for the first time. Even more startling was the fact that newcomers like HTC and ZTE had ascended into the top five ranking. Nokia though remained the second largest maker of cellphones after Samsung. The phenomenal partnership of two indus-try titans could win a sliver of share in the mobile platform wars and brought both Microsoft and Nokia to a dilemma: succeed or else.

Amid all this uncertainty, when Nokia announced in December 2012 that it would sell its headquarters in suburban Helsinki, but would continue to hold the office by leasing it, it was seen as yet another tremor in Nokia's shaky world. The Finnish firm had moved out of central Helsinki in 1996 to this iconic building on the edge of the sea in the neighboring town of Espoo. The 540,000-square-foot facility overlooking the Gulf of Finland included rich Scandinavian interior-design features and a massive employee cafeteria with hardwood floors. The wings of this building—also known as Nokia House—were connected by glass-enclosed bridges. It was completed in three phases; the final portion of the building was finished in 2001 when Nokia still dominated the global handset market.

Nokia headquarters, also dubbed as NoHo, was ten minutes of drive away from the center of Helsinki. Nokia had sold its strikingly beautiful headquarters to software consultancy firm Exilion for US$222 million but had agreed to lease it back at a lower price. To be fair, it was pretty common for large corporations to extract capital out of their real estate when needed for their core business. However, this large glass-and-wood building along the waterfront in Espoo, Finland symbolized Nokia's ascending to the top of the wireless world. And the fact that transaction came at a time when the mobile phone giant was fighting for survival carried an important symbolism.

Next up, Nokia moved out of the building to make way for Microsoft after the sale of is Devices & Services unit to American software giant was complete in April 2014. Nokia staff moved to another building nearby in Karaportti;

Nokia owned this building. Petra Soderling, a former Nokia employee, told *CNET* that she felt emotional when her Facebook stream started filling with pictures of the glowing blue Nokia sign being taken down at Nokia House in Espoo, Finland, and being replaced with a white Microsoft logo.

Nokia had introduced Finland on the world map to a large number of people across the globe. It had been the country's largest employer and most valuable company. In 2000, for instance, Nokia contributed 4 percent of the country's gross domestic product. However, by 2012, that number was estimated to be around 0.2 percent. Moreover, at its peak, in 2002, Nokia contributed 21 percent of all of Finland's corporate tax revenue. In its heydays, Nokia used to employ 1 percent of the Finnish workforce, but now that was in the region of 0.2 percent. Around 40 percent of all the money spent on R&D in Finland came from Nokia in 2009; again, after the sale of Nokia's mobile devices unit, the new estimated figure was 17 percent.

In the final analysis, Nokia leadership admitted that its partnership with Microsoft was a mistake. Nokia's gamble to side with Microsoft looked right on the paper because it brought together the best of Microsoft's software prowess and Nokia's hardware excellence and product engineering. However, it didn't work in reality because the two companies were going against the flow set by Android and iOS mobile platforms. And the fact that the fates of Microsoft and Nokia were intertwined in this epic smartphone battle made the whole affair far more complicated. In 2013, it was apparent that both Microsoft and Nokia were moribund players in the smartphone game, and they both had to get their acts together. They were both running out of time and goodwill despite deep pockets, organizational power, and brand recognition.

Microsoft was desperate to regain some of the ground from Apple and Google in the smartphone realm. But if its major partner, Nokia, was being seen as backing away, that put Microsoft back at square one. So, at this peculiar crossroads, for Microsoft, it was apparently more viable to own technology versus being a partner since Microsoft wasn't in control of what Nokia could do. The market dynamics essentially forced Microsoft to buy Nokia's mobile phone business. That, in turn, gave Nokia another stab at reinventing the company through a renewed focus on data-centric wireless infrastructure market.

ELOP'S SCORECARD

In 2010, then-chairman Jorma Ollila flew to the United States on a three-day trip to interview five potential candidates for the top job at Nokia. Stephen Elop was second choice in the search for a new CEO but was subsequently hired because the primary candidate withdrew himself from the selection process due to personal reasons. Elop impressed the Nokia chairman as a good salesman and a decisive corporate executive. Ollila was also impressed with Elop's experience in the software industry and his Finnish-style directness. Elop, a native of Ancaster, Ontario, became the first non-Finn to be the chief executive of Nokia.

According to Ollila, the Nokia board was initially leaning toward replacing Olli-Pekka Kallasvuo with an internal candidate. But the gravity of Nokia's predicament demanded a decisive leadership and Elop was known in the industry for his tremendous energy and drive. The Canadian executive who built his résumé at the U.S. technology companies carried this amazing ability to create a transparent and fast process. His strengths included a deep knowledge of the dynamics of the software industry. In fact, software expertise was the main reason Elop made to the list of candidates in the first place. As the head of Microsoft's US$19 billion Office business, Elop had run one of the world's largest and most profitable software operations.

Elop grew up in Ontario and graduated in computer engineering and management from McMaster University in Hamilton, Canada. His first major spell was a six-year stint as chief information officer of restaurant chain Boston Chicken. In 1998, after McDonald's bought over Boston Chicken, he joined Macromedia which produced web design tools for Apple developers. It was Macromedia who made the Flash video software that powered the rise of YouTube, and Dreamweaver, the software widely used for building websites. Elop held a number of senior positions at the San Francisco–based software house and eventually ascended to the CEO role in 2005.

Three months later, he brokered Macromedia's US$3.4 billion acquisition to the leading graphics software firm Adobe Systems Inc. When it became clear that he was not going to become the next Adobe CEO, he jumped ship to networking gear maker Juniper Networks Inc. Here, he served as chief operating officer and was buoyed by assurances that he would soon replace CEO Scott Kriens. But just days before he was to be named chief executive, Steve Ballmer called to recruit him to Microsoft. During all these years, Elop developed a reputation for embracing challenges and resolving internal conflicts.

In retrospect, Nokia was looking for someone who could bring the caliber of Steve Jobs to the struggling European giant. Or a leader like Louis Gerstner, who helped revive IBM by getting all the pieces to work together. But Elop was no visionary; he was known to be a clever operator. He ran Nokia more like a penny-pinching CFO than a visionary CEO. A bumbling strategy became the hallmark of Elop's managerial style, and that didn't go well with the Nokia rank-and-file. Charles Fitzgerald, Microsoft's former general manager of Platform Strategy, was quite harsh when he took an insider scoop at Elop's profile: "His resume is that of a short-tenured opportunist who has left little mark on his employers, except off course Nokia where he presided over the company's collapse and ultimate exit from the mobile handset business."

The critics blamed Elop of selling Nokia's soul to Microsoft. Some pundits even said "he just doesn't get telecom" and that telecom was so different than the software industry from where Elop had descended. There was a rallying cry that Nokia should not have hired a former Microsoft executive as its new chief. Elop had apparently pulled the company into a tighter embrace with his former employer, setting Nokia on a highly risky and uncertain path with no guarantee of success. Elop was certain that he would be accused of being a double agent for his former company as he clinched the landmark software deal with Microsoft.

Some of the ex-Nokians branded him the "world's worst CEO." Mobile industry analyst and former Nokian Tomi Ahonen was notably blunt in pointing out that the former Microsoft man had presided over a massive decline at Nokia. He had suggested at one stage that Elop's removal would be best for the Finnish company. He quoted the fact that Nokia's smartphone unit sold 28 million devices per quarter and had 29 percent market share when Elop took over reins of the wireless concern on September 21, 2010. At that time, Ahonen added, Nokia was twice as big as Apple and three times bigger than Samsung.

Nokia's first non-Finnish CEO was tasked with rebuilding Finland's equivalent—in status and prominence—of 1980s era General Motors, a monolith at the height of its powers, yet in an undeniable decline. Elop came on board at a time when Nokia's reputation and market share were already tanking. And there were all indications that things could get really bad at Nokia before they started getting better. Under his watch, the company began to embark upon a years-long reorganization that would initially leave it with a much smaller share of the mobile handset market.

Nokia under Elop was no more an incumbent in the mobile space. It was now the underdog, and that was a weird position for the Finnish wireless firm. Still, it wasn't copying anyone else's market strategy because Elop, the architect of Nokia's comeback strategy, knew very well that smartphone fix would only come through a sound and well-thought process. The book has attempted to line up all the progress that Elop accomplished in the make-over of a less Finnish Nokia. The notes from Nokia diaries chronicled in this book show that Elop with his decisive leadership style made some headway.

First, Elop brought speed and a sense of urgency to an organization that had grown fat and complacent while sitting on top of a huge pile of profits over the years. He made many tough choices, including major layoffs, selling office space at Nokia headquarters and shedding aging software technologies like Symbian. Those calls prevented an even worse outcome and gave the Finnish company a badly needed sense of urgency. Elop pushed Nokia to move more quickly than ever before. Second, Nokia's triumph in the mapping and location domains was a testament that the Finnish telecom player was finally able to sort out the software world. Third, as mentioned above, Nokia had discarded its market-leader mentality. The Finland–based company was now a lot smaller and people inside Nokia were working together in a better cohesion.

Elop had even made headway on improving distribution in the United States, where the company had little presence in the preceding years. He had also reinforced engineering perspective to the development of smartphones, and he was more intimately involved with the process than former executives. However, it was a make-or-break moment for Nokia and Elop's challenge was two-fold. First and foremost, he had to preserve Nokia's soul that laid in its mobile phone usability and its powerful brand. On top of that, he had to ensure an effective execution of Nokia's Windows Phone strategy that was persistent as well as receptive to adjustments down the way.

In a nutshell, the real challenge for Nokia continued to be the smartphone segment. The steady loss of the smartphone market meant that Nokia would end up competing with Chinese vendors and potentially experience a slow death. In the end, Elop failed to stop the company's declining smartphone market share and left Nokia as a work in progress. Furthermore, he proved to be too much of an American-style corporate executive for the Finnish company that had cultivated a low-key corporate culture.

THE END OF AN ICON?

The prologue of the book had asked if Nokia could reinvent itself like IBM did during the 1990s. In retrospect, IBM had remained stagnant for too long before it regained the old stature in the computing industry. Nokia, by contrast, took the wake-up call way too early and knew very well that it had to fight back to remain relevant in a highly competitive mobile industry. Finland's stumbling mobile industry giant could also afford a lot of pain amid the cushion provided by its other profitable business units. Nokia's three non-handset business operations—wireless gear manufacturing, mapping services, and patents portfolio—could make up for its handset business and keep Nokia in the league of large technology companies. Nokia for sure was not facing the risk of imminent closure.

The industry might not want to write off Nokia. The Finnish electronics manufacturer had been on the ropes before. Nokia had come very close to selling its then-fledgling mobile phone business back in the 1990s. Boston Consulting Group had done a thorough assessment of Nokia's business in 1991 and concluded that the Finnish company wouldn't be able to compete with Motorola and the Japanese mobile handset makers. At that time, Nokia looked more like an obscure Korean chaebol and was largely controlled by Finnish banks. Many of its stakeholders were state agencies. They tried selling the entire company to Ericsson, but the Swedes saw Nokia's television business as being too big and too risky. Nokia's television sets business was huge with too many basic models and too much R&D spread over too many locations.

A year later, in 1992, Siemens wanted to buy Nokia's cellular systems unit, but Jorma Ollila turned them down. By that time, the board had taken a chance on Ollila, who began turning the company into a mobile phone giant. Nokia persisted with mobile and continued investing in cutthroat mobile technology throughout the 1990s and beyond. Meanwhile, in 1996, Ollila sold Nokia's television manufacturing business which had weighed heavily on the company's bottom line. About two decades later, the Finnish industry was at a similar crossroads, aiming to reinvent itself in the burgeoning wireless infrastructure business.

Nokia's foray into the booming LTE equipment business seemed akin to its 1990s thrust into digital cellular communications. And just the way the European GSM project had vindicated Nokia's mobile vision back then, a major breakthrough was now knocking the LTE bandwagon door during the mid-2010s. Mobile operators—Nokia Networks' primary customers—were aiming their turbo-speed LTE networks to serve the Internet of Things a.k.a. machine-to-machine communications (M2M). The Internet of Things was the manifestation of an open, global network connecting people, data and machines. Pundits called it the biggest opportunity in the history of technology business. By 2020, billions of things—from clothes to body sensors to tracking tags—were forecast to be connected to mobile networks. And that could consume 1,000 times as much data as mobile gadgets of mid-2010s.

The mid-2010s had also marked the beginning of the era of connected wearable devices that were now being embedded into daily life objects like cars and homes. And all this was happening at the cross-section of mobility, cloud computing and big data. Google Glass was a classical example of such an amalgam of innovative new technologies that promised to define the next phase of mobile computing revolution. The connected wearables like Glass and smartwatches were an extension of the smartphone, and they worked as satellite devices that amassed useful data or relayed notifications from a primary mobile device. Industry watchers envisaged these wearable systems to become the primary means of accessing the web by 2020.

The leading network equipment makers like Cisco and Ericsson clearly seemed excited on the possibility of having 50 billion devices connected to the web by 2020. Now the new Nokia was also aiming for the next web and this future roadmap was written all over the press release that announced the appointment of Rajeev Suri as the new company head. "Nokia believes that over the next 10 years billions of connected devices will converge into intelligent and programmable systems that will have the potential to improve lives in a vast number of areas: time and availability, transportation and resource consumption, learning and work, health and wellness, and many more." Even Suri's first statement on taking over the top job seemed to reinforce that future vision. "The world of technology is on the verge of a change that we believe will be as profound as the creation of the Internet."

Nokia had unwrapped a new era, and it was all set to reinvent itself as a mobile infrastructure powerhouse amid exciting new opportunities. In the final analysis, it seemed evident that the end of the road for Nokia—the mobile industry icon—had been greatly exaggerated. The still-giant but often-maligned mobile company was very much in the game. Things could change rapidly in the technology world. ...And the wireless market was still a wide-open field.

Nokia's headquarters in Espoo, Finland symbolized the company's rise to the top of the wireless world. There are no dark corridors at this seaside construction near Helsinki. The building, transparent to the sky and water, is covered by a thermodynamically efficient layer of 26,000 plates of glass.

Image: Nokia

"Company cultures are like country cultures. Never try to change one. Try, instead, to work with what you've got."

Management guru Peter Drucker's famous quote was probably the antithesis of Stephen Elop's tenure at Nokia. The first non-Finn head of Nokia attempted to transform the company culture in a bid to revive the Finnish phone giant. Instead, he ended up with a broken organization. Some people in Finland called him "Stephen Eflop."

NOTES

Prologue

Bobbie Johnson, "Everything that's wrong with Nokia in one blog post," *Gigaom*, April 13, 2012.

Bolaji Ojo, "Will Nokia Rise Again?" *EE Times*, May 26, 2011.

Eric Zeman, "Nokia At Risk Of Joining RIM And Palm," *Information Week*, June 14, 2012.

Henry Blodget, "Nokia Implodes... Taking Microsoft's Mobile Dreams Down With It," *Business Insider*, June 15, 2012.

"Kodak files for bankruptcy protection," *The Economist*, January 19, 2012.

Om Malik, "Why Kodak's bankruptcy should scare Nokia," *Gigaom*, January 19, 2012.

Chapter 1

Kevin J. O'Brien, "Nokia's New Chief Faces Culture of Complacency," *The New York Times*, September 26, 2010.

Majeed Ahmad, "Smartphone: Mobile Revolution at the Crossroads of Communications, Computing and Consumer Electronics," *CreateSpace*, December 16, 2011.

Walter S. Mossberg, "Nokia Steps Into Race For 'Communicator' With a Weak Start," *The Wall Street Journal*.

Chapter 2

Andrew Orlowski, "The rise and fall of the great Finnish phonemaker," *The Register*, September 6, 2013.

"A Finnish fable," *The Economist*, October 14, 2000.

Kevin J. O'Brien, "Nokia's New Chief Faces Culture of Complacency," *The New York Times*, September 26, 2010.

Natasha Lomas, "Innovate Or Die: Nokia's Long-Drawn-Out Decline," *TechCrunch*, December 31, 2012.

Chapter 3

Brian Proffitt, "Hey! You've Got Your Moblin in My Maemo!" *IT World*, February 17, 2010.

Christopher Mims, "Here Comes the First Real Alternative to iPhone and Android," *Quartz*, December 2, 2012.

Ewan Spence, "Finland's Jolla Will Be The Ferrari Of The Smartphone World," *Forbes*, July 21, 2012.

Junko Yoshida, "Chasing Apple, Nokia calls up all developers," *EE Times*, April 29, 2009.

Priya Ganapati, "Intel's MeeGo OS Runs Into Rough Weather," *Wired*, October 8, 2010.

Priya Ganapati, "Symbian OS Is Broken. Can It Be Fixed?" *Wired*, October 25, 2010.

Chapter 4

Junko Yoshida, "Analysis: Is Nokia under pressure?" *EE Times*, May 23, 2008.

Junko Yoshida, "Opinion: Nokia vs. Nokia," *EE Times*, May 1, 2009.

Junko Yoshida, "Too late to save Nokia?" *EE Times*, December 6, 2012.

Matthew Lynn, "How Nokia Fell From Grace," *Bloomberg Businessweek*, September 15, 2010.

"N-Gage History," *Wikipedia*, February 16, 2011.

Peter Kirwan, "Nokia's last stand," *Wired*, May 9, 2012.

Richard C. Morais, "Bloody but unbowed," *Forbes*, May 14, 2001.

Rick Merritt, "Opinion: Mobile Industry will eat Nokia's lunch," *EE Times*, September 1, 2009.

Chapter 5

Bobbie Johnson, "How do you solve a problem like Nokia?" *Gigaom*, July 20, 2012.

Dan Rowinski, "Nokia Versus Android: Death by a Thousand Cuts," *ReadWrite*, June 14, 2012.

Frederic Lardinois, "Nokia's Richard Kerris: People Won't Remember Our Troubles By Next Spring," *TechCrunch*, July 8, 2012.

Junko Yoshida, "Too late to save Nokia?" *EE Times*, December 6, 2012.

Kevin J. O'Brien, "Together, Nokia and Microsoft Renew a Push in Smartphones," *The New York Times*, February 11, 2011.

Nick Wingfield and Christopher Lawton, "Nokia's Flirtations Put the Fear of Google Into Microsoft," *The Wall Street Journal*, February 18, 2011.

Peter Burrows, "Stephen Elop's Nokia Adventure," *Bloomberg Businessweek*, June 2, 2011.

Ryan Kim, "Symbian is alive and kicking, for now," *Gigaom*, December 30, 2011.

Seth Weintraub, "Nokia-Microsoft deal is good news for Android," *Fortune*, February 11, 2011.

Vlad Savov, "Competition is king: why Nokia and Microsoft are the perfect match," *The Verge*, April 9, 2012.

Chapter 6

Dan Rowinski, "Nokia has one job: Drive the growth of Windows Phone," *ReadWrite*, November 13, 2013.

"Difference Engine: True to its image," *The Economist*, December 3, 2012.

Heidi Lemmetyinen, "Zooming in on Nokia PureView," *Conversations by Nokia*, February 29, 2012.

Kevin C. Tofel, "Nokia's Lumia Transition Is Complete. Will It Pay Off?" *Gigaom*, September 6, 2012.

Michal Lev-Ram, "Can Nokia's Lumia take on Apple and Google?" *Fortune*, January 10, 2012.

Michal Lev-Ram, "Can the Lumia smartphones save Nokia?" *Fortune*, September 5, 2012.

Roger Cheng, "Farewell Nokia: The rise and fall of a mobile pioneer," *CNET*, April 25, 2014.

Tom Simonite, "A Comeback Phone, Hampered by a Lack of Apps," *MIT Technology Review*, April 3, 2012.

Chapter 7

Chuck Jones, "The Market Essentially Forced Microsoft To Buy Nokia," *Forbes*, September 3, 2013.

Ewan Spence, "Could We See A Finnish Smartphone From Nokia In 2016?" *Forbes*, September 5, 2013.

Joe Wilcox, "Has Stephen Elop doomed Nokia?" *Beta News*, June 27, 2011.

Kevin C. Tofel, "Samsung keeps smartphone sales crown, Nokia drops from top 5," *Gigaom*, October 26, 2012.

Kevin J. O'Brien, "Hoping for Turnaround, Nokia Bets on Windows 8," *The New York Times*, September 2, 2012.

Mark Jacobstein, "Every Phone a Smart Phone," *Fortune*, September 14, 2009.

Natasha Lomas, "Nokia Puts Social Networking Knobs On ~$62 Series 40 Asha 205, 205 Dual-SIM Qwerty Handsets," *TechCrunch*, November 26, 2012.

Nilay Patel, "There will never be another Nokia smartphone," *The Verge*, September 3, 2013.

Quentin Hardy, "From Nokia, an Executive Who Knows the Difficulties at hand," *The New York Times*, September 3, 2013.

Shira Ovide and Sven Grundberg, "Microsoft to Buy Nokia Mobile Business in $7 Billion Deal," *The Wall Street Journal*, September 3, 2013.

Tim Stevens, "For Nokia, helping the competition find its way is good business," *Engadget*, September 28, 2012.

Tristan Louis, "Why Apple Should Acquire Nokia," *TNL.net*, October 6, 2012.

Chapter 8

David Zax, "The Rise of Nokia's Maps," *MIT Technology Review*, October 5, 2012.

Juhana Rossi and Ian Edmondson, "Nokia Pays $2.21 Billion for Siemens Stake in SNS," *The Wall Street Journal*, July 1, 2013.

Junko Yoshida, "Nokia, not Google, sees itself reshaping the Internet," *EE Times*, February 11, 2008.

Junko Yoshida, "Nokia's naked ambition: Moving beyond cellphones," *EE Times*, October 23, 2008.

Kevin Fitchard, "As Nokia Siemens Shrinks the 4G Network, Its Prospects Grow," *Gigaom*, October 19, 2012.

Mark Scott, "Digital Mapping May Be Nokia's Hidden Jewel," *The New York Times*, April 24, 2014.

Rachel Metz, "Nokia Unveils a Map Service that Lives in the Cloud," *MIT Technology Review*, November 13, 2012.

Stuart Dredge, "Nokia doubles down on maps and location-based services," *The Guardian*, November 14, 2012.

Chapter 9

Craig Wilson, "Can Nokia be saved?" *TechCentral*, June 2, 2011.

Ewan Spence, "Nokia Had To Choose Windows Phone, Can They Make The Strategy Work?" *Forbes*, June 16, 2012.

Haydn Shaughnessy, "If Nokia Disappears, What a Case Study That Would Make," *Forbes*, October 18, 2012.

Juhana Rossi, "Elop Was Second Choice as Nokia CEO," *The Wall Street Journal*, October 17, 2013.

Kevin J. O'Brien, "Nokia Has Some Good News After Two Years of Gloom," *The New York Times*, January 10, 2013.

Roger Cheng, "Nokia on the edge: Inside an icon's fight for survival, *CNET*, December 18, 2012.

Tom Krazit, "Nokia's Elop: We can do things with Windows Phone that Microsoft can't," *Gigaom*, October 2, 2012.

INDEX

ABOUT THE AUTHOR

Majeed Ahmad is former Editor-in-Chief of *EE Times Asia*, a sister publication of *EE Times*. While being the Editor-in-Chief at Global Sources, a Hong Kong–based publishing house, he also spearheaded magazines relating to electronic components, consumer electronics, and computer, security and telecom products.

This is his sixth book on wireless and smartphones. His other five book titles are *Smartphone, The Next Web of 50 Billion Devices, Mobile Commerce 2.0, Age of Mobile Data,* and *Essential 4G Guide.*

He is currently associated with a number of technology publications as a contributing writer and Editor-at-Large. He has been a technology and trade journalist for more than 18 years.